Practical Guide
to Environmental
Impact Assessment

Environmental Engineering Books

ALDRICH • *Pollution Prevention Economics: Financial Impacts on Business and Industry*

AMERICAN WATER WORKS ASSOCIATION • *Water Quality and Treatment*

BAKER, HERSON • *Bioremediation*

BRUNNER • *Hazardous Waste Incineration, Second Edition*

CALLAHAN, GREEN • *Hazardous Solvent Source Reduction*

CASCIO, WOODSIDE, MITCHELL • *ISO 14000 Guide: The New International Environmental Management Standards*

CHOPEY • *Environmental Engineering in the Process Plant*

COOKSON • *Bioremediation Engineering: Design and Application*

CORBITT • *Standard Handbook of Environmental Engineering*

CURRAN • *Environmental Life-Cycle Assessment*

FIKSEL • *Design for Environments: Creating Eco-efficient Products & Processes*

FREEMAN • *Hazardous Waste Minimization*

FREEMAN • *Industrial Pollution Prevention Handbook*

FREEMAN • *Standard Handbook of Hazardous Waste Treatment and Disposal*

HARRIS, HARVEY • *Hazardous Chemicals and the Right to Know*

HARRISON • *Environmental, Health, and Safety Auditing Handbook, Second Edition*

HAYS, GOBBELL, GENICK • *Indoor Air Quality: Solutions and Strategies*

JAIN, URBAN, STACEY, BALBACH • *Environmental Impact Assessment*

KALETSKY • *OSHA Inspections: Preparation and Response*

KOLLURU • *Environmental Strategies Handbook*

KOLLURU • *Risk Assessment and Management Handbook for Environmental Health and Safety Professionals*

KREITH • *Handbook of Solid Waste Management*

LEVIN, GEALT • *Biotreatment of Industrial and Hazardous Waste*

LUND • *The McGraw-Hill Recycling Handbook*

MAJUMDAR • *Regulatory Requirements of Hazardous Materials*

ROSSITER • *Waste Minimization Through Process Design*

SELDNER, COETHRAL • *Environmental Decision Making for Engineering and Business Managers*

SMALLWOOD • *Solvent Recovery Handbook*

WILLIG • *Environmental TQM, Second Edition*

Practical Guide to Environmental Impact Assessment

Betty Bowers Marriott

McGraw-Hill

New York   San Francisco   Washington, D.C.   Auckland   Bogotá
Caracas   Lisbon   London   Madrid   Mexico City   Milan
Montreal   New Delhi   San Juan   Singapore
Sydney   Tokyo   Toronto

Library of Congress Cataloging-in-Publication Data

Marriott, Betty B.
    Environmental impact assessment : a practical guide / by Betty B.
Marriott.
        p.   cm.
    Includes bibliographical references and index.
    ISBN 0-07-040410-0
    1. Environmental impact analysis—United States.   I. Title.
TD194.65.M37   1997
333.7'14—dc21   96-39003
                                                           CIP

McGraw-Hill

A Division of The McGraw-Hill Companies

            12 13 IBT/IBT   0 1 0 9

ISBN 0-07-040410-0

*The sponsoring editor for this book was Zoe G. Foundotos, the editing
supervisor was Caroline R. Levine, and the production supervisor was
Pamela A. Pelton. This book was set in Century Schoolbook by Victoria
Khavkina of McGraw-Hill's Professional Book Group composition unit.*

Printed and Bound by Integrated Book Technology

McGraw-Hill books are available at special quantity discounts to use
as premiums and sales promotions, or for use in corporate training programs. For more information, please write to the Director of Special
Sales, McGraw-Hill, 11 West 19th Street, New York, NY 10011. Or contact your local bookstore.

 This book is printed on recycled, acid-free paper containing a
minimum of 50% recycled, de-inked fiber.

Contents

Introduction

Environmental impact assessment is an integral component of decisions made every day on proposed projects, plans, and actions. The National Environmental Policy Act (NEPA) requires that a statement of environmental impacts of proposed projects be prepared as part of the development process of federally funded projects. Several states also have environmental protection laws in effect which are applicable even if federal funding is not involved. These laws, and numerous related environmental regulations and statutes at federal, state, and local levels, affect almost every construction project and planning activity in the United States and in many other parts of the world. An understanding of the essentials of the NEPA process and requirements, and those of other environmental statutes and permits, is important to those who pursue a career in this field, those who construct or plan projects or communities' facilities, public officials, and even the general public who will certainly be affected at some level by the environmental impact process.

The purpose of this book is to improve the quality and efficiency of environmental impact assessment, and the subsequent worthy contribution of the completed analysis to the decision-making process. This book provides, in one location, the fundamental essentials of impact assessment in all impact categories.

Chapters 1 and 2 explain the foundation for environmental studies; critical terms; basic documents and content; and the required processing for the three thresholds of documentation. The remainder of the text goes beyond *what* needs to be achieved to equally focus on *how* to conduct high-quality, efficient analysis of potential project or action impacts.

Chapter 3 examines the importance of early and continuing agency and public involvement in the impact assessment process. The text recommends productive techniques to solicit and incorporate meaningful input and to avoid conflicts or time delays. Chapter 4 describes the extreme importance of alternatives development and scrutiny to the impact analysis and to the overall success of a project, plan, or action. It emphasizes the critical and continuous connection of alternatives refinement and impact analysis.

The next 17 chapters are what makes this book different from many others. Rather than just list the types of impacts that need to be investigated in an environmental assessment, this book explains the essential elements of impact analysis within each category. The text is full of examples, recommendations, and guidance that will become immediately useful in the practical, competent, and time-efficient conduct of environmental analyses. Each impact category chapter identifies applicable legislative content and processing requirements, permits, and required agency coordination. Methodologies are described and distinguished by type or magnitude of the project being assessed. Common errors to avoid are revealed. No text can present exactly what needs to be included in an environmental impact assessment or particular disciplinary areas of impact, because seldom are any two proposed actions exactly the same. By presenting numerous examples of the types of impacts that may occur with various types of projects, the text invokes the thought process to ensure sensible and effective identification of possible impacts that may otherwise be overlooked. Mitigation measures in successful use are discussed.

The last chapter focuses on the importance of productively using results of environmental assessment in the evaluation of proposed alternatives. Methodologies to facilitate input critical to the decision-making process are recommended and described.

This text provides practical guidance, discussions, and thought-provoking examples. Direction is given on the efficient compliance with established legislation, executive orders, regulations, and guidelines. The process required to successfully complete quality and timely environmental documents that are focused on real issues, and thus add to the decision-making procedure, is described, with tips on common mistakes to avoid, on cost- and time-saving measures, and on productive coordination with agencies and the public to ensure that major issues are addressed early in the studies. The reader will learn the importance of alternatives development and the fact that the process of evaluating and enhancing specific features of proposed alternatives is a dynamic, not stagnant, ongoing action throughout the environmental assessment.

In addition to providing guidance on practical preparation and processing of documents to comply with legislative and regulatory requirements, the text—especially those chapters on specific disciplinary areas of potential impact—meshes the fundamental, practical requirements of environmental impact assessment with underlying principles and theories. The reader thus will understand the reasons why certain changes produced by proposed projects or actions should be included in the assessment of potential adverse or beneficial impacts. This feature of the book expands its usefulness beyond the "cookbook" legal requirements of large federal or state-funded projects to effective environmental assessment at all levels of local, state, and federal decision making.

The environmental assessment process includes insurance that decisions are not made without consideration of all interests. Through the use of multidisciplinary teams, the analysis of impacts is a result of interaction of team members with expertise in many fields of study. By requiring public and agency review and input, the process ensures that all elements of value to the human environment are considered. The various agencies established within the United States are charged with the protection and management of particular areas of the nation's resources. Through their review and input, these agencies, with jurisdictional and technical expertise involvement, ensure a representation of all public interest resources into the decision-making process.

The information presented here is not meant to be a detailed study of every complex regulation or methodology, which can change with time and also take extensive study and experience to master. Rather, this approach to the environmental process explains the basics of required documentation levels; identification of analysis requirements in the disciplines that should be evaluated; the public and agency participation process; and the timing and processing of environmental activities within the context of overall project development.

For this reason, detailed descriptions of all regulations and current complex technical methodologies are not given. Numerous agencies have developed specific and detailed guidelines to be followed on projects under their jurisdictions. Often a multidisciplinary team conducts the environmental assessment and analyses for any particular project. Each team member is educated and experienced in his or her particular area of expertise. This book is targeted to those at the beginning of a future specialization and those wishing to obtain an overall understanding of the content and process of environmental impact assessment and its underlying principles.

The information here is not, however, just for the beginner. It is equally targeted, and provides a valuable resource, for those established experts and for federal and state agency personnel specializing in a particular disciplinary area of resource protection. In this sense, the text provides critical guidance in areas not within the particular expertise of the experienced professional. The knowledge gained can enhance the agency coordination process that is key to successful environmental impact assessment and mitigation by providing a foundation understanding of other impact areas not within the specialization of the experienced expert. For beginning professionals to seasoned agency, public administration, engineering, or planning personnel, this text will be a constantly valuable reference.

Environmental impact assessment is fundamentally an analysis of the changes produced by a particular project or action. The earth which we humans inhabit with all other living and physical entities is constantly in a state of change to continually reach a balance of unanimity within the community. Because of the interaction and interrelationship of all the earth's elements, every change produces associated changes. Natural occurrences, such as earthquakes, floods, hurricanes, droughts, fires, or volcanic eruptions, can produce catastrophic and immediate changes. Most of the changes caused by living organisms, however, occur at an extremely slow rate. In contrast, changes produced by the human species can occur rapidly and abruptly. For every change, the process of reaching a new balance begins again.

Sometimes the results of human-produced changes are not beneficial and leave a surrounding environment that we neither wished for nor expected. Although the proposed project or action has good intent and solves or addresses an identified problem, the resultant ramifications may be a degradation of the human environment beyond the benefits provided by the proposed project or action. This is the underlying principle and purpose of environmental impact assessment.

An environmental impact assessment identifies those associated changes produced by a proposed project or action. It evaluates the degree of these changes, or impacts, on the basis of short-term or long-term, adverse or beneficial, direct and indirect effects. It ensures that (1) alternatives are developed that minimize adverse environmental impacts and (2) mitigation measures to reduce associated impacts are included in the planning process. It promotes a comparative assessment of all proposed alternatives, including the alternative of doing nothing (the no-action or no-build alternative), to yield an understanding of the cost/benefit tradeoffs and thus lead to better decisions.

1

Foundation

Certain concepts of environmental impact assessment are established in the language of the National Environmental Policy Act and implementing regulations. These concepts are fundamental to all impact assessments (Fig. 1.1).

1.1 National Environmental Policy Act

The National Environmental Policy Act (NEPA) of 1969, as amended (42 U.S.C. 4321–4347), set forth requirements for agencies of the federal government in Title I and established the Council on Environmental Quality (CEQ) in Title II. Among the most significant features of the law in Section 102(2) are the requirements for federal agencies to use a systematic, interdisciplinary approach to ensure integrated use of environmental arts in planning and decision making which may have an impact on the human environment; to develop procedures to ensure that environmental amenities and values are given appropriate consideration in decision making, along with economic and technical considerations; and to

> Include in every recommendation or report on proposals for legislation and other major Federal actions significantly affecting the quality of the human environment, a detailed statement by the responsible official on—
>
> (1) The environmental impact of the proposed action,
> (2) Any adverse environmental effects which cannot be avoided should the proposal be implemented,
> (3) Alternatives to the proposed action,
> (4) The relationship between local short-term uses of man's environment and the maintenance and enhancement of long-term productivity, and

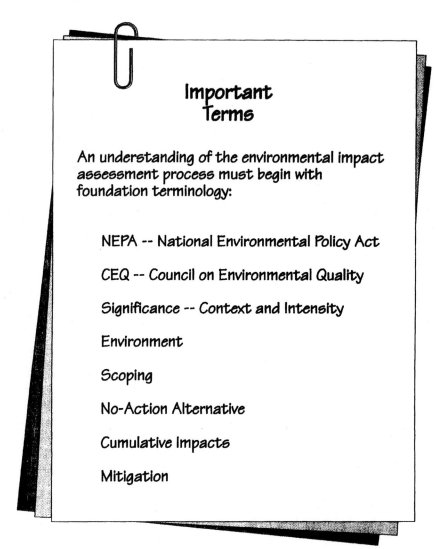

Important Terms

An understanding of the environmental impact assessment process must begin with foundation terminology:

NEPA -- National Environmental Policy Act

CEQ -- Council on Environmental Quality

Significance -- Context and Intensity

Environment

Scoping

No-Action Alternative

Cumulative Impacts

Mitigation

Figure 1.1 Important terms.

(5) Any irreversible and irretrievable commitments of resources which would be involved in the proposed action should it be implemented.

The above Section 102(2)(C) is the part of the legislation often referenced in what has become known as the Environmental Impact Statement (EIS).

1.2 Council on Environmental Quality Regulations

The CEQ issued guidelines in 1970 and revised guidelines in 1973 for implementation of Section 102(2)(C) of the law. The guidelines established a process, defined required environmental documents, and set forth recommendations for public review and involvement. Because the guidelines were intended to be nondiscretionary standards for agency decision making, but were viewed by some agencies and courts as advisory only, President Carter issued Executive Order 11991 on May 24, 1977, directing the CEQ to issue regulations.

In 1978, the guidelines became regulations (40 CFR Parts 1500–1508), and efforts were made to reduce paperwork, reduce delay, promote better decisions, and focus environmental studies on issues and impacts that were relevant. The CEQ's regulations became binding on all federal government agencies, replaced some 70 different sets of agency regulations, and provided uniform standards applicable throughout the federal government for conducting environmental reviews.

All federal agencies have developed guidelines and regulations for implementation of the CEQ regulations. Some agencies have very detailed guidelines, while others are more general in nature and follow the basic guidance of the CEQ. Although all agencies are implementing the same CEQ regulations, the specific agency guidelines and regulations can differ substantially from one another. The environmental analyst must be aware that each agency (and usually state) has individual established methods of complying with CEQ regulations. Normally there is a good reason for these differences in procedure; each agency's guidelines are molded to the activities, structure, and policies characteristic of that particular agency. The environmental professional who works with numerous agencies and within numerous states should be forewarned that there frequently will be tangible and abundant differences in the "right" way to comply with CEQ regulations (Fig. 1.2).

1.3 Role in Better Decisions

The language of the Executive Order reveals some of the early problems that had evolved in the 1970s, by mandating that the regulations be

> ...designed to make the environmental impact statement more useful to decision makers and the public; and to reduce paperwork and the accumulation of extraneous background data, in order to emphasize the need to focus on real environmental issues and alternatives.

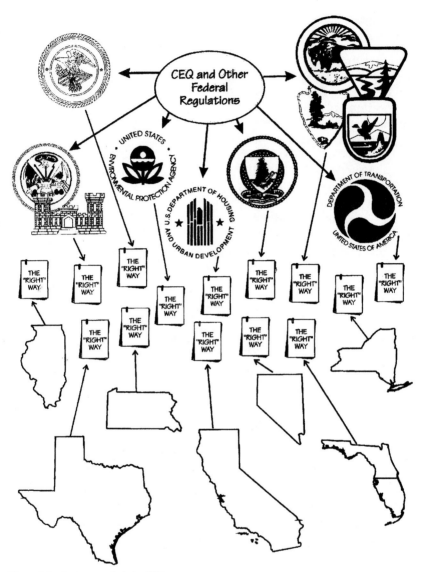

Figure 1.2 Agency and state differences.

Whereas the guidelines were limited to subsection (C) of Section 102(2) of NEPA—the requirement for environmental impact statements—the regulations included all the Section 102(2) provisions for agency planning and decision making. The emphasis on better decision making was a response to a problem of the environmental impact statement's becoming an end in itself and failing to establish the link between what is learned through the NEPA process and how the information can contribute to better agency decisions. Important information was being generated through the NEPA process, but it was resulting in large, encyclopedic environmental documents and was not playing a large enough role in concise comparison of the pros and cons of various proposed alternatives for a particular project or action.

A multidisciplinary approach to environmental impact analysis is critical to the decision-making process and to the equal consideration of all areas of potential impact, when the tradeoffs of particular alternatives are evaluated. Therefore, the professionals assessing impacts within a particular area of impact, such as natural resources, air quality, or neighborhood effects, must be educated and qualified within the disciplinary area they are assessing. The regulations addressed another emerging problem of environmental documents sometimes being prepared by a single individual, or a group of individuals of similar narrow range of qualifications, by requiring a List of Preparers in environmental documents that clearly states the qualifications of the individuals responsible for preparation of the various disciplinary sections of the environmental document.

1.4 Significance

Several important basics to environmental impact assessment are established in the CEQ regulations. The determination of *significance* is defined in terms of context and intensity. *Context* refers to the geographical setting of a proposed project or action. When a proposed shopping center is evaluated, for example, the context for the determination of significance is the immediate setting and the general community or area of influence, not the United States or the world as a whole.

Intensity refers to the severity of impact. Criteria included in the CEQ regulations are

- The degree to which the proposed action affects public health or safety
- Presence of unique characteristics in the geographic setting or area, such as cultural resources, parklands, wetlands, ecologically critical areas, or wild and scenic rivers

- The degree to which the effects are likely to be highly controversial
- The degree to which the action would establish a precedent for future actions with significant effects
- The degree to which possible effects are highly uncertain or involve risks
- The degree of effect on sites listed in the National Register of Historic Places
- The degree of effect on threatened or endangered species or their habitats
- Whether the action conflicts with other federal, state, or local laws or requirements

Most federal agencies and many state agencies have defined potential projects under their jurisdiction which normally produce significant effects on the human environment and thus normally require preparation of a Draft and a Final Environmental Impact Statement.

The criteria of significance have expanded over time as a result of legislation, guidelines, court decisions, and other influences. Use of the word *significant* has become somewhat controversial; some agencies now refuse to include it in environmental documents because it assumes a judgment, and other agencies insist that every possible impact be tagged *significant* or *nonsignificant*. The identification of significant impacts often becomes critical to an agency's commitment to provide mitigation for an expected impact.

1.5 Environment

The phrases *effect on the human environment* and *environmental assessment* often are misunderstood by many to mean natural resource effects. In fact, the range of considered environmental impacts is very comprehensive and includes artificial components of the environment as well as natural. Disciplinary areas of possible effects include ecological, aesthetic, historic, cultural, economic, social, and health. More detailed listings of possible impact categories are given in the discussion of the Environmental Impact Statement (EIS) format and in later chapters dealing with the content and procedures for environmental impact assessment documents.

1.6 Scoping

The CEQ regulations established the requirement of *scoping* at the early stages of environmental impact assessment and thereby reinforced a commitment to an organized, systematic program of agency and public participation in the environmental process. Scoping refers to early coordination with interested and affected agencies and with the public.

The scoping process, conducted early in the environmental impact assessment process, identifies (1) important issues and concerns, (2) areas not of concern for a particular project or action, and (3) other legislative or regulatory requirements. The scoping process, discussed in greater detail in Chap. 3, thus establishes the scope of additional studies, assists in staffing and scheduling of study activities, and promotes the compliance with all applicable legislative requirements within an integrated study and document.

1.7   Alternatives

The purpose of an environmental impact analysis, or evaluation, is to comparatively evaluate alternative courses of action. The range of alternatives considered must include a no-action, or no-build, alternative and other reasonable courses of action. Although early environmental studies considered alternatives as alternatives *to the proposed action,* current studies generally recognize the importance of giving all alternatives equal status in the level or degree of analysis and design development.

1.8   Indirect and Cumulative Impacts

An interdisciplinary approach is used to consider and assess environmental impacts. The analysis considers potential consequences which are long-term and short-term; direct and indirect, or secondary; individual and cumulative; beneficial and adverse.

Indirect, or secondary, effects are those that may occur removed in distance or time from the actual proposed project. An example is the construction of a major employment center, which may have direct effects related to aesthetics in the area, traffic at nearby intersections, removal of natural vegetation, or interference with natural waterways. Additional employment opportunities in the location, however, may prompt additional housing or commercial uses to support employees. Potential impacts of this housing or additional business activity would then be a secondary, or indirect, effect of the construction of the employment center and should be evaluated to the best extent possible in the environmental analysis.

Cumulative impacts occur in those situations where individual projects or actions may not have a significant effect, but when combined with other projects or actions, the individual project's incremental contribution of adversity may cause an overall adverse cumulative effect. Cumulative effects may be *additive* or *interactive*. Additive effects are the same sources of impact affecting the same resource of the environment. If one project removes a service, such as day care, from a neighborhood, the impact may not be considered critical because other similar services exist. But if other projects are proposed which remove these similar services, then a significant adverse cumulative effect may result. This is an example of an additive effect. Piecemeal physical destruction of wetlands is another example of an additive cumulative effect.

An interactive effect is created when differing sources of impact affect the same environmental resource. A small stream relocation may be determined to not cause significant effects of erosion and sedimentation; but if other projects, even those outside of the jurisdiction of the sponsoring agency, add toxic pollutants to the same stream, then the contribution of sediment due to the first project may be considered to have a significant adverse cumulative effect. This is an example of an interactive cumulative effect on the water quality and aquatic life of the stream.

The environmental analyst must remember that indirect and cumulative effects are removed in distance or time. The assessment should include a review of future actions in a particular area and the probable impacts. For example, the Bureau of Land Management requires preparation of a Reasonably Foreseeable Future Action (RFFA) or Reasonably Foreseeable Development (RFD) scenario for the project area to assist in assessment of possible indirect and cumulative effects.

Evaluations of indirect and cumulative effects have always been a requirement of the CEQ regulations for implementation of NEPA. The importance of, and attention given to, these types of impacts has intensified in recent years, particularly in highly developed, densely populated areas of the country.

1.9   Defining the Action or Project

The action or project being proposed must be carefully evaluated to determine the appropriateness of its definition for environmental impact assessment and documentation. If this sounds complex, an example will show why this required definition of proposed action is important. A major, complex project is proposed which will likely have significant effects on the human environment. Instead of considering the complete project or action in a single analysis or document, the proposed project or action is split into small pieces and each piece is analyzed in a separate document. The results of such a piecemeal approach will likely yield several pieces of project with minimal effects and will not disclose the total project or action effect.

Proposed projects or actions evaluated within environmental studies must be shown to have independent utilities (have uses on their own with no other required projects) and to not foreclose or preclude future options. Highway projects are good examples of the independent utility concept. A highway project evaluated in an environmental document should have a logical beginning and end and should effectively respond to an identified need. A highway agency would not, for example, propose environmental impact studies of a bridge over a river in the middle of nowhere with no connections for separate evaluation in an environmental document. Obviously, whatever is proposed to connect to the bridge must be implemented to make the bridge a viable project, and therefore it needs to be included in the description of the proposed project undergoing environmental analysis.

In determining the scope of an Environmental Impact Statement, agencies must consider three types of actions (other than unconnected single actions): connected actions, cumulative actions, and similar actions (CEQ 1989). Connected actions

- Automatically trigger other actions that may require Environmental Impact Statements (EISs)
- Cannot or will not proceed unless other actions are taken previously or simultaneously
- Are interdependent parts of a larger action and depend on the larger action for their justification

Similar actions have similarities that provide a basis for evaluating their environmental consequences together, such as common timing or geography.

1.10   Additional Required Considerations

In addition to the analysis of the environmental impacts of the proposed alternatives, the CEQ regulations state that the environmental-consequences discussion should include

- Any adverse environmental effects which cannot be avoided
- The relationship between short-term uses of the human environment and the maintenance and enhancement of long-term productivity
- Any irreversible or irretrievable commitments of resources

The way in which these requirements have been addressed over the years often has been awkward. Preparers of environmental documents sometimes are confused about the proper location of these discussions within a particular document, and, in the case of the last two requirements, about the content and intent of the discussion.

Some agency outlines for EISs give these areas full section status, usually at the end of the document, following the environmental-consequences section. Other agencies add headings at the end of, but within, the environmental-consequences section. In other cases, it is assumed that these types of considerations have been generally included within the overall analysis for each subject area of potential effect, and no separate heading or status is given at all.

Unavoidable adverse effects are best listed in the summary of an environmental document. The impacts to be highlighted are those which cannot be mitigated to acceptable levels. If a separate mitigation summary or report is prepared, a listing of unavoidable adverse effects also is contained in that discussion.

The relationship between short-term uses and the maintenance and enhancement of long-term productivity is usually a very generalized discussion. Short-term uses may include loss of existing resources for land clearance, disruption of neighborhoods, or benefits such as improved transportation, more efficient energy use, improved utility services, generation of economic stimuli, and creation of jobs.

Long-term productivity includes natural resources of the existing environment, the environmental quality of life, or perhaps a discussion of foreclosure of future options for use of land. This discussion is basically an identification of the tradeoffs involved. Often area or regional planning documents are a good source of information regarding the relationship of the proposed project or action to long-term productivity.

The assessment of irreversible and irretrievable commitments of resources customarily is approached in a general, rather than specific, manner. Types of resources would include the actual materials, labor, and funds expended for a construction project. Some natural resource losses are not likely reversible, even in a long-term time frame. Commitments of resources also may include increased public infrastructure and services, particularly for projects which promote intensified development.

1.11 Mitigation

All possible measures to mitigate potential impacts should be included in the proposed action. The CEQ regulations define mitigation to include

1. *Avoiding* the impact

2. *Minimizing* the impact by limiting the degree or magnitude of the action

3. *Rectifying* the impact by repairing, rehabilitating, or restoring the affected environment

4. *Reducing* or eliminating the impact over time

5. *Compensating* for the impact by replacing or providing substitute resources or environments

The study of, and commitment to, mitigation measures have become an extensive and meaningful part of the environmental impact assessment process. In many circumstances, potentially adverse impacts can be avoided or mitigated to acceptable levels through careful design and implementation of appropriate measures or techniques to reduce the severity of the effects. Erosion control measures, noise walls, relocation assistance, and construction of replacement habitat or wetlands are examples of successful mitigation techniques.

1.12 Related Environmental Requirements

In addition to NEPA and its CEQ implementation regulations, several other environmental directives have evolved since the early 1970s. These directives are contained in Executive Orders, Code of Federal Regulations (CFR) regulations, legislation, department orders, Technical Advisories, policy memoranda, and numerous other documents. Frequently, environmental directives relate to particular resources, or impact categories, such as historic resources, air quality, scenic rivers, or floodplains. These separate legislative and regulatory requirements are often the reason for inclusion of particular discussions within an environmental impact document. For resources subject to such requirements specified in statute, regulation, or executive order, the environmental analyst must conduct studies to demonstrate no impact as well as possible impacts.

During the past 20 years, states also have implemented environmental legislation and regulations. Often these state regulations closely follow the corresponding federal agency. For example, a state department of transportation may use the Federal Highway Administration/Federal Transit Administration's guidelines. Some states have peculiar specific requirements which must be integrated with the federal guidelines on a project with shared federal and state funding or which must be followed alone on projects with only state or local funding. The names for the various thresholds of documentation also vary among states and often differ from those established in the CEQ regulations.

1.13   Litigation

Throughout the history of the NEPA and related environmental legislation and regulations, courts have influenced the environmental impact assessment process through case law interpretation. Often these cases deal with definitions within the law or regulations, such as *significantly* affecting the quality of the human environment, *cumulative* impacts, and *all reasonable alternatives*. The study of environmental law case history and its influence is a subject not addressed in this text. Most court decisions, however, are reflected in the current guidance offered by particular agencies or within guidelines on implementation of specific state environmental laws. It is possible to have an understanding of the environmental impact assessment process without an understanding and knowledge of all the influences that court decisions have had in shaping particular guidelines.

2

Environmental Documents and Processing

This chapter discusses the three basic thresholds of environmental analysis documentation and the required processing for each level. The applicability of each document and the required content are explained. Processing procedures differ with each threshold level of assessment documentation.

2.1 Preliminary Overview Assessments

Often the first step in an environmental impact assessment is a preliminary overview of the proposed project alternatives and locations. Several steps are included in the overview. First, the project alternatives' characteristics must be reviewed. Is the project a building, highway, park, or land-use plan? What are the characteristics of the setting? Is the potentially affected area urban or rural, natural or made by humans?

The purpose of the preliminary assessment is to identify the potential for significant environmental impacts of the initial set of alternatives. Results then function to refine the alternatives and to determine the appropriate subsequent environmental documentation and process. The three thresholds of environmental processing and documentation are Categorical Exclusion (CE), Environmental Assessment/Finding of No Significant Impact (EA/FONSI), and Draft and Final Environmental Impact Statements (DEIS and FEIS) (Fig. 2.1). An initial assessment determines which documentation and processing methodology should be followed: (1) a Categorical Exclusion for minor projects and actions requiring no environmental

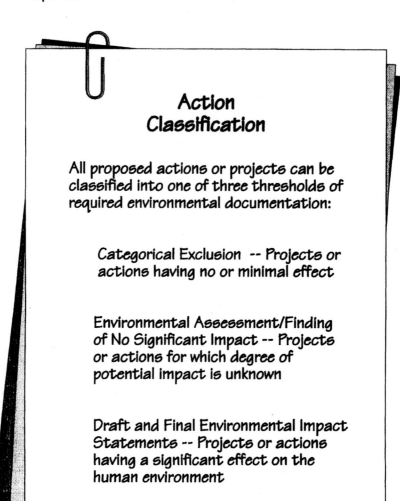

Figure 2.1 Action classification.

clearance, (2) an EIS for projects and actions significantly affecting the quality of the human environment, and (3) an EA/FONSI for everything in between.

Another purpose of an initial environmental overview is its contribution to the development of the project, or action, alternatives. Results from an early overview assessment can provide important "fatal flaw" information to the designers or sponsors of the proposed project or action. By providing this information early in the project development process, the initial overview can lead to better designs or plans that avoid obvious pitfalls which may arise later in the planning stages. Problems that arise late in the planning stages of any project or action normally are significantly more costly and time-consuming to address.

Several agencies and states have developed forms for conducting initial environmental overviews. Although the forms may have different names, such as Environmental Assessment Forms (New York State) or Environmental Checklists (California), they have all been developed to identify potentially significant impacts and issues early in the project planning process. Through a series of questions, the form leads the evaluator to simple answers, such as yes, no, or maybe, on the potential for specific significant impacts in the various disciplines to be assessed. Based on the answers to these questions, the forms lead to a decision on the subsequent project processing requirements. A few examples of the types of questions included in an initial assessment overview, in areas of potential physical, biological, social, and economic impacts, follow: Will the proposal either directly of indirectly

- Modify the channel or a river or stream?
- Reduce the critical habitat of any unique, threatened, or endangered species?
- Divide or disrupt an established community?
- Require the displacement of businesses or farms?

An immediately obvious observation about the types of questions listed in the above examples is that the answers to the questions are often not available until after a more detailed inventory of existing setting and a more thorough analysis of potential impacts have been made. Therefore, answers to questions are often *maybe* for any project of substance. Environmental assessment forms do help to determine that a Categorical Exclusion is appropriate for a very minor type of project or action, and sometimes they assist in focusing subsequent environmental studies for more complicated projects. Use of a formal form, however, often is not required. The experience and qualifications of the professional conducting the initial overview will normally be sufficient to recognize the potential for significant impacts, or fatal flaws, in a particular project.

The best methodology to use is to proceed systematically through all the categories of potential environmental effects and, based on the available preliminary information about the proposed project or action and its environmental setting, determine whether the potential exists for impacts which normally would require preparation of an EIS. If it is uncertain whether *major, significant* environmental effects would occur, an EA/FONSI approach should be followed. The scoping process also will assist in this determination between an EA/FONSI or EIS direction of study.

The CEQ regulations of 1978 required each federal agency to develop procedures to implement and supplement the regulations. Each agency's procedures were to include specific criteria for, and identification of, those typical classes of action which normally (1) do require Environmental Impact Statements, (2) do not require either an Environmental Impact Statement or an Environmental Assessment (Categorical Exclusions), and (3) require Environmental Assessments but not necessarily Environmental Impact Statements. In compliance with this directive, most federal agencies and many states have an established list of types of projects under their jurisdiction which normally require the three thresholds of evaluation, documentation, and processing. For example, each agency within the Department of the Interior has prepared, as appendices to the departmental manual, information on

1. NEPA responsibilities

2. Guidance to applicants

3. Major actions normally requiring an EIS

4. Categorical Exclusions

If the particular project or action is clearly defined on one of these three lists of types of actions, then a preliminary environmental overview becomes more of a task in focusing subsequent studies and improving design or other characteristics of proposed alternatives, rather than determining the appropriate threshold of environmental documentation. Often, however, a particular project does not exactly meet all the conditions of the examples on these established lists. It is in these circumstances that the initial environmental overview becomes most critical.

One of the most important contributions of an initial overview assessment is the early input of environmental considerations to the design or development of the project, action, or plan. If coordination is efficient among the various members of the team for the project or action, the information provided by an initial overview can lead to better projects with fewer potential environmental impacts. These "least environmentally damaging" alternatives are then the ones evaluated in the subsequent detailed environmental studies and public and agency review process.

2.2 Categorical Exclusion

To reduce paperwork, the CEQ regulations of 1978 established the Categorical Exclusion (CE). In doing so, the regulations recognized that many agencies have minor projects and activities which normally do not have an effect on the environment. Each federal agency has established criteria and, in most cases, a list of typical projects for which environmental evaluation normally is not required. Additionally, some agencies have published lists of exceptions to categorical exclusion, for projects which may otherwise be excluded but have particular characteristics that warrant more detailed studies.

An example of the CE process is that used by the Department of the Interior, Bureau of Land Management (BLM) (1988). The BLM procedures for conducting a Categorical Exclusion review include these steps:

1. Ensure conformance with the land-use plan.

2. Review the DOI departmental and BLM lists of Categorical Exclusions to determine if the proposed action falls into one of the listed categories.

3. Check proposed action against the list of exceptions to determine if any apply.

If exceptions apply, an Environmental Assessment (EA) or Environmental Impact Statement is required. If no exceptions apply, the proposed action may be categorically excluded.

Procedures for processing a Categorical Exclusion vary among agencies and among states for state or locally funded projects. Typically, a brief documentation is required which explains how and why the particular proposed action meets the criteria, or is a named typical project, for a Categorical Exclusion. Explanatory text should accompany the basic document if the project does not exactly meet the definition of any particular listed typical project, but has been determined to produce no effect on the environment. Based on the degree of required additional documentation for these "exceptions" to the defined, listed projects, some agencies have established multiple levels of documentation required for Categorical Exclusions.

Categorical Exclusion documentation remains within the agency and its files. No formal circulation to other agencies or public involvement procedures are required. Coordination with responsible agencies and/or the public may be included, however, in the documentation of applicability of the criteria for Categorical Exclusion to a particular unlisted project.

If the initial evaluation and overview of the project result in uncertainty for classification as a Categorical Exclusion, the agency or responsible official may decide to proceed with preparation of an Environmental Assessment.

2.3   Environmental Assessment/Finding of No Significant Impact (EA/FONSI)

An EA/FONSI is the second threshold of environmental analysis and documentation. The level of detailed studies required is greater than that for a Categorical Exclusion, but less than that for an Environmental Impact Statement. The degree of analysis and documentation conducted with an EA can vary significantly according to the complexity of the proposed project or action being assessed.

2.3.1   When an Environmental Assessment is appropriate

An Environmental Assessment—capital E, capital A, as compared with a generic (lowercase) environmental assessment or analysis—is a document prepared when a project does not readily fit into either a Categorical Exclusion or an Environmental Impact Statement category. The potential significance of environmental impact is not clearly established. The purpose of an EA is to provide sufficient evidence to determine whether the proposed project or action will require a full Environmental Impact Statement or a Finding of No Significant Impact. An EA is the selected threshold of documentation when impacts will occur but will be minor or can be successfully mitigated to acceptable levels.

2.3.2   Format and content

The format for an EA is established in CEQ regulations and, as with other required documents, is further refined by the sponsoring agency or state guidelines and regulations. An EA should have the following sections at a minimum:

I. Need for proposed action

II. Description of alternatives

III. Environmental impacts

IV. List of agencies and persons consulted

The areas of potential impact to be evaluated follow those also used for an EIS. The difference between an EIS and an EA is that the number of relevant issues for an EA project normally is fewer than those of a project for which an EIS is prepared. Early coordination, or scoping, should be conducted to (1) assist in determining which aspects of the proposed action have potential for social, economic, or environmental impact; (2) identify alternatives; (3) specify possible measures to mitigate potential impacts; and (4) identify other environmental review and consultation requirements which should be met concurrently with the EA.

In an EA, the description of the existing resources and environment of the affected area or site is contained in the same section as the discussion of environmental impacts. Therefore, for example, the subsection on water quality would begin with a description of area streams, groundwater resources, water supply system, and other features relevant to the understanding of impact. Then, in the same section, the discussion would include the impacts of the project on these resources. This format approach differs from that of an EIS, which separates the existing, or *affected,* environment and the environmental impacts, or *consequences,* into two separate main sections.

An EA normally results in a FONSI, but can be revised to an EIS if studies reveal that significant impacts will occur. If the document is finalized as an Environmental Assessment, the subsequent document is a FONSI. There are no "draft" and "final" Environmental Assessments as there are in the EIS threshold of documentation.

2.3.3 Processing

When completed, the availability of the EA for review is announced to the public and to interested federal, state, and local agencies. The notice of availability includes a brief description of the proposed project or action and its environmental impacts. All interested parties are invited to submit comments, in writing, within 30 days.

Although a formal public hearing is not required, most agencies will conduct one if requested by an agency or by the public. There also is no formal requirement to circulate an EA to other agencies, organizations, or officials; but most sponsoring agencies do circulate and offer the opportunity for comment to appropriate federal and state agencies and locally affected municipalities or regions. The EA must be made available for public and agency review for 30 days prior to issuing a FONSI.

2.3.4 Finding of No Significant Impact

After the 30-day review and comment period, if no significant adverse impacts are identified, the sponsoring agency may issue a FONSI. The FONSI is normally brief because it includes the EA or incorporates it by reference. Often only a few pages long, the FONSI can be attached to the front of an EA to ensure that the EA remains with it as a reference.

The FONSI is the formal determination of the sponsoring agency that the proposed project or action will not have a significant adverse environmental impact and that an EIS will not be prepared. It briefly describes the reasons for this conclusion. Notice of availability of the FONSI is given to the public and to affected federal, state, and local agencies.

Some agencies require—and it is always a good policy to prepare—a decision record at the conclusion of an EA/FONSI process. The record includes

- Identification of the selected alternative and the rationale for the decision

- The FONSI and attached EA

- Reasons why an EIS was not required

- A compliance and monitoring plan for any mitigation commitments made as part of the decision

2.4   Draft and Final Environmental Impact Statements

The most detailed procedure for analyzing potential environmental impacts of alternatives of a proposed project or action is the Environmental Impact Statement process. The use of the word *draft* in a Draft Environmental Impact Statement (DEIS) should not be confused to mean an internal draft or preliminary document not seen by others. The DEIS contains the final results of environmental studies of proposed alternatives which are available for public and agency review. The DEIS is a "draft" because it compares all proposed alternatives and is the document upon which the decision to proceed with any particular alternative is made. The DEIS also is the tool through which public and agency input is incorporated into this decision-making process. The Final Environmental Impact Statement (FEIS) documents which alternative has been selected as the preferred and the reasons for that selection.

Certain EIS processing documents and activities are specifically required by the CEQ regulations, while others are agency-specific requirements, are common practice, or are needed to prepare the documents noted in the CEQ regulations. Figure 2.2*a* and *b* shows a recommended EIS process flow diagram for a fairly complex project. Not all the steps or documents contained in the figure, however, are necessarily required for every project. For example, the CEQ regulations require public participation and a public hearing, but not necessarily the public meetings noted in the flow diagram. Similarly, separate methodology and technical reports may not be appropriate for

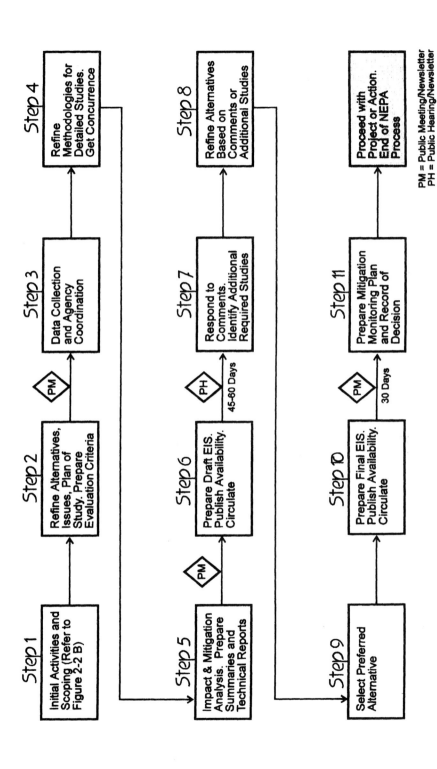

Step 1
Initial Activities and Scoping (Refer to Figure 2-2 B)

Step 2
Refine Alternatives, Issues, Plan of Study. Prepare Evaluation Criteria

PM

Step 3
Data Collection and Agency Coordination

Step 4
Refine Methodologies for Detailed Studies. Get Concurrence

Step 5
Impact & Mitigation Analysis. Prepare Summaries and Technical Reports

PM

Step 6
Prepare Draft EIS. Publish Availability. Circulate

PH
45-60 Days

Step 7
Respond to Comments. Identify Additional Required Studies

Step 8
Refine Alternatives Based on Comments or Additional Studies

Step 9
Select Preferred Alternative

Step 10
Prepare Final EIS. Publish Availability. Circulate

PM
30 Days

Step 11
Prepare Mitigation Monitoring Plan and Record of Decision

Proceed with Project or Action. End of NEPA Process

PM = Public Meeting/Newsletter
PH = Public Hearing/Newsletter

Figure 2.2 (*a*) A recommended process for projects and actions requiring an Environmental Impact Statement.

25

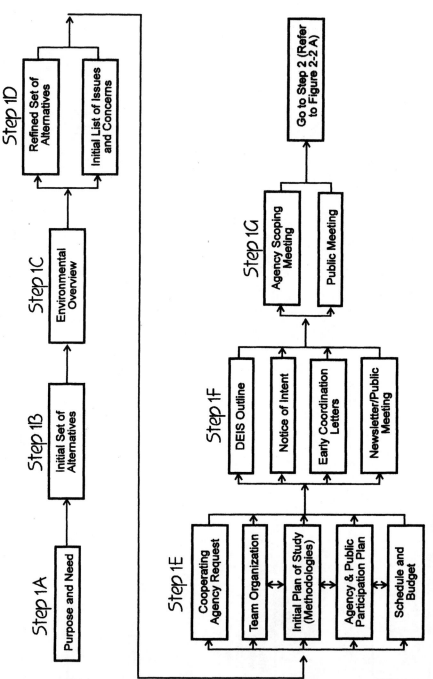

Figure 2.2 (b) A recommended process for projects and actions requiring an Environmental Impact Statement—expanded step 1.

every project. In Fig. 2.2*b*, scoping, Notice of Intent, and Cooperating Agency Request are specifically noted as requirements in the regulations, whereas a written Plan of Study and an Environmental Overview are not. These tasks, however, are often necessary to prepare the Notice of Intent.

2.4.1   When an Environmental Impact Statement is appropriate

An Environmental Impact Statement is prepared for projects or actions which will have a significant effect on the human environment. As noted previously, the CEQ regulations contain guidance on determining the significance of impacts, and federal agencies have issued regulations or guidelines defining what actions within their jurisdiction normally require an EIS.

2.4.2   Use of previously prepared Environmental Impact Statements

Prior to the actual beginning of a Draft and Final EIS process, existing environmental documents should be reviewed. Under certain circumstances, if the proposed project or action is covered within a previously prepared environmental document and meets all the relevant criteria for using that document, then no new EIS is required. This situation often arises in agencies, but environmental documents prepared by other agencies also can be used. If a previously prepared NEPA document fully covers a proposed action and no additional analysis is necessary, that determination should be properly documented in the project files.

2.4.3   Notice of Intent

The first formal step in the EIS process is the preparation of the Notice of Intent, a brief document announcing the intent of the sponsoring agency to prepare an EIS for its proposed project or action. In addition to a description of the project, the Notice of Intent contains a description of the scoping process to be used and an announcement of the formal scoping meeting, if one is to be held. A formal document in the process, the Notice of Intent is published in the *Federal Register.*

As noted in Fig. 2.2*a* and *b*, several other project initiation activities need to occur prior to the preparation of the Notice of Intent, including the development and screening of alternatives. The initial set of alternatives cannot be developed without first defining the purpose and need for the proposed project or action. A formulated plan for public participation and agency scoping also must be developed prior to the Notice of Intent.

2.4.4 Determination of lead and cooperating agencies

Other activities taking place at the initiation of EIS studies are the determinations of lead agency and cooperating agencies. These activities are usually very straightforward. The lead agency is the agency supervising the preparation of the EIS. Cooperating agencies are those which have jurisdiction by law, usually through permitting, or special expertise in any specific environmental impact associated with the proposed project. State or local agencies also can be designated as cooperating agencies upon their request.

Although designation of cooperating agencies is not usually a major effort, it is an activity that must be formally conducted and should not be overlooked. It is the responsibility of the lead agency to formally request, in writing, other agencies to become cooperating agencies. If another agency has jurisdiction by law, that agency must become a cooperating agency. Agencies with specialized expertise may, upon their decision, choose to become cooperating agencies.

Although the language of the CEQ regulations implies that a cooperating agency should play a relatively significant role in the environmental analysis in its area of expertise, including making staff available and using its own funds, such involvement commonly does not occur.

2.4.5 Scoping

The next step in the EIS process is early coordination and scoping. These areas are discussed in greater detail in Chap. 3 and consist of early contacts with interested agencies, with the purpose of defining areas of concern and focusing the subsequent environmental studies on relevant issues. The scoping and early coordination process is extremely important in shaping the level of analysis, staff requirements, and schedule of all subsequent activities in the preparation and processing of Draft and Final EISs.

2.4.6 Development of alternatives

Another activity that occurs early in the process is the initial development of alternatives and the subsequent screening of alternatives for a final list of those to be considered in the EIS studies. Alternatives are discussed in detail in Chap. 4. The initial set of alternatives will most likely be developed by the engineers, architects, or planners designing the proposed project prior to scoping. The screening process by which the initial list is reduced to a set of reasonable alternatives considered in the analysis process should have the input of an initial environmental overview for fatal flaws and the results and input from scoping and early coordination with interested agencies. These potential environmental concerns are then considered with other criteria, such as costs, engineering factors, and community acceptance in reducing the number of alternatives to be considered. Depending on the type of proposed project, the alternatives may consist of alternative sites or locations for development or facilities.

2.4.7 Impact analysis

After the scope of study has been determined and the reasonable alternatives defined, the next activity is the analysis of environmental impacts. Activities begin with a description of the existing environment and the assembly of relevant information and data. The evaluation and analysis of degree of impact follow. Considered impacts must include direct and indirect effects, cumulative effects, and long- and short-term effects, as discussed in Chap. 1. In the analysis process, potential mitigation measures are developed and explored. Content and procedures for analysis in the various disciplines of potential impacts are included in Chaps. 5 through 21.

The preparation of separate methodologies and technical reports supporting the DEIS has become common practice, to limit the discussion in Sections III and IV of the DEIS. These technical reports are prepared according to the discipline area and contain the detailed information on existing conditions, methodologies, analysis, and results. The technical reports are then summarized in the DEIS.

Technical reports supporting a DEIS can be as few as 4 or 5 or as many as 20. Each agency has its own preference. At a minimum, technical reports are normally prepared for

- Socioeconomic impacts, to include community impacts, land use, economic impacts, visual effects, relocations, traffic, and pedestrian and bicycle travel
- Natural resources, to include water quality, vegetation, wildlife, scenic rivers, floodplains, wetlands, and coastal zones
- Air quality
- Noise

Separate reports also may be required in discipline-specific legislation or regulations. Examples are reports for historic and archaeological resources in compliance with the National Historic Preservation Act and its guidelines, sole-source aquifer reports, wetland delineations, and wildlife surveys. Although an attempt should always be made to include as many of these requirements as possible in the DEIS for simultaneous circulation and review, specific timing or coordination requirements may dictate the preparation of separate reports.

Some agencies are more comfortable with separate technical reports for each impact area, such as water quality, vegetation, wildlife, visual impacts, floodplains, and land use. The decision to prepare a technical report should be based on the need to have a separate document to hold a large amount of material not needed by the average reader of the DEIS for an understanding of the setting and impacts of the proposed project.

Technical reports serve the purpose of providing supporting data to the information summarized in the DEIS. The DEIS serves the purpose of providing a comparative summary of adverse and beneficial results among the reasonable alternatives being considered to result in a decision on the preferred alternative.

2.4.8 Format and content of a DEIS

Following the completion of analysis, the DEIS is prepared. The basic format of an EIS is established within the NEPA, the CEQ regulations, and individual agency guidelines. At a minimum, a DEIS should have the following components:

Cover sheet

Summary

Table of Contents

   I. Purpose of and Need for Proposed Action

  II. Alternatives

 III. Affected Environment

 IV. Environmental Consequences

List of Preparers

List of Agencies, Organizations, and Persons to Whom Copies of the DEIS are Sent

Index

Appendices

The CEQ regulations prescribe the use of concise, clear language in an EIS and the limitation of data and information to those that are relevant. A limit on the number of pages also is recommended. These directions were prompted by the tendency for EISs to become voluminous documents full of extraneous material not relevant to the decision at hand (i.e., to select a preferred alternative). The author of an

EIS should always critically review all the information to determine whether its inclusion in an EIS is necessary and relevant.

The level of analysis also should be directly dictated by the magnitude of expected impact, as determined by the initial overview studies and results of the scoping and early coordination efforts.

The first main section of an EIS covers the *Purpose and Need* for the Proposed Action. This chapter is a very critical component to the document as a whole, and its importance should not be underestimated. A brief description of the proposed action should be given. Then the need for the action should be thoroughly described. Often this can be accomplished by reference to planning documents or other long-range plans that dictate a need for the action. The discussion usually begins with a description of the deficiencies or problems in the current situation, i.e., without the proposed action.

Goals and objectives for the project should be established that directly relate to the deficiency or problem that the proposed project is attempting to address. These should be outlined in Section I. This section of the EIS becomes the basis upon which the alternatives and analysis of environmental effects are ultimately considered.

Although documented in the DEIS or EA, the definition of purpose and need is considered much earlier in the process than the preparation of the environmental document. As noted in Fig. 2.2*b*, it is one of the initial project activities because the development of alternatives is based on the identified needs and deficiencies. Evaluation criteria and parameters to be used to determine how well each proposed alternative responds to the purpose and need also should be developed early in the process, prior to actually conducting impact analysis.

The *Alternatives* section of the DEIS describes all reasonable alternatives being considered, including the proposed action and the no-action alternatives. Alternatives are discussed in detail in Chap. 4. The no-action, or no-build, alternative is defined as existing and future conditions without any improvements to correct the deficiencies identified in Section I (Purpose and Need). It should include, however, other planned projects for the area.

Agencies differ in opinions on the progress that other projects must have made to be included in the definition of the no-build alternative. Some do not allow inclusion of any proposed or planned projects unless they are currently under construction. Some agency guidelines call for inclusion of only planned *and funded* projects, and other agencies include anything on adopted planning documents.

The no-action alternative has the purpose of establishing a basis upon which all other proposed build alternatives are evaluated. Therefore, its definition should be agreed upon by all parties, after careful consideration. The definition of the no-action alternative can cause the results of environmental impact analyses to vary significantly since many impacts are measured by the degree of change compared with the no-build alternative.

The Alternatives section of the DEIS contains a detailed description of each proposed alternative, including physical characteristics, operating features, costs, schedule, description of the construction process, and all other relevant features of the proposed action. Readers should have a comprehensive understanding of what is being proposed and how the proposed alternatives differ from one another and from the no-build alternative.

The *Affected Environment* section of the DEIS contains information on the existing setting. Although the organization and format vary, the following areas should be included, if relevant to the proposed project and its impacts:

Land use and zoning

Social and neighborhood characteristics
- Demographic characteristics
- Housing
- Travel patterns
- Stability
- Pedestrian and bicycle travel
- Community facilities and services (fire and police, hospitals, schools, churches, day care, etc.)
- Recreational facilities

Economic factors
- Taxes
- Existing business community
- Proposed developments

Traffic and transportation

Energy

Historic and archaeological resources

Visual resources

Air quality

Noise levels

Geology and soils (including farmland)

Environmental health and public safety (hazardous wastes)

Water resources
- Groundwater
- Surface water
- Water supply and wastewater systems

■ Wild and scenic rivers

Wetlands

Floodplains and coastal zones

Vegetation and wildlife

The most common error in the preparation of a DEIS is the inclusion of entirely too much information in this section. This is not always the fault of the preparer, however. Frequently agencies, now used to seeing an abundance of descriptive information, will request detailed information in their area of expertise. Motives for this request are understandable—to expand the data base of information in their field at no cost to the particular agency itself. The environmental analyst must remain focused, however, on the particular data needed for determining the impact of the project or action being evaluated.

The level of detail in Section III is directly driven by the anticipated magnitude of impact. It is accepted that certain basic descriptive information must be included to establish a complete setting of social, natural, and economic features of the proposed action. Too often, however, this section has become an encyclopedic accumulation of information not necessary to understand the degree of impact subsequently presented under the same headings in Section IV, Environmental Consequences. A good rule of thumb is that Section III of the DEIS should be the smallest of the four major sections.

The *Environmental Consequences* section of the DEIS contains the results of the assessment of impacts. It can be organized by impact category or by alternative; the usual format is by impact category. The list of categories should follow those used in Section III, Affected Environment.

As noted previously, this section should focus on relevant environmental issues and impacts. Some areas of potential effect must be included, however, regardless of expected impact. Resources protected by statute, regulation, or executive order must be addressed in all environmental documents. When such protected resources do not exist within the area or will not be affected, the EIS must document that the resource was considered in compliance with the applicable regulation, and statements must be made as to why the resource will not be affected or why the regulation does not apply. For example, the DEIS may contain statements such as "There are no designated wild and scenic rivers in the project area" or "No listed threatened or endangered species or habitat is located within the area of potential adverse effect" or "None of the proposed project alternative sites are located within a coastal zone or floodplain." These negative declarations are necessary because of the specific laws and regulations applicable to these resources. Acceptable methods for ruling out possible impacts are discussed within Chaps. 5 through 21.

Although it is not required by the CEQ regulations, a DEIS should contain another section following the Environmental Consequences section. Frequently entitled *Coordination and Consultation,* this section documents all coordination and consultation conducted throughout the project with agencies and the public. It describes the scoping process and any public information meetings held on the project. Agency coordination is further discussed in Chap. 3.

2.4.9   DEIS processing

When the DEIS is completed, it is circulated to all federal, state, and local agencies having jurisdictional or expertise interest in the proposed action. In some cases the summary of the DEIS can be circulated in lieu of the entire document. Notices are placed in newspapers to notify the public of the availability of the DEIS and at what locations in the community it may be reviewed. A DEIS can be purchased by any member of the public for the printing cost of the document. Availability of the DEIS also is published in the *Federal Register* through the U.S. Environmental Protection Agency (EPA).

All notices invite comments on the DEIS and give the due date for comments. The usual review period for a DEIS is 45 days. In certain circumstances this can be shortened to 30 days, but often in practice it extends to 60 days. Some regulations, such as those of the Department of the Interior, require a 60-day minimum review period for DEISs prepared by their agencies. Individual review agencies also can request an extension of time to prepare comments.

The public notice of availability of the DEIS and the cover letter circulating the document to agencies and officials customarily include information on the public hearing. A public hearing is required on a DEIS and must be held within the 45-day public and agency review period, but not sooner than 30 days after availability of the DEIS is announced.

Following the public hearing and the end of the public review period, all comments received are reviewed and evaluated. Any required additional analysis is identified and conducted. Alternatives and mitigation measures may be refined based on comments received. Responses to each substantive comment are prepared.

Based on the review of all comments received and the results of any additional studies, the sponsoring agency then selects the preferred alternative. This selection process should be a systematic evaluation procedure, and it is discussed more fully in Chap. 22. The process then continues to the preparation of the Final EIS.

2.4.10   Final Environmental Impact Statement

The FEIS documents the choice of the preferred alternative. It consists of the DEIS with modifications. In some cases, where minor changes are required, an abbreviated form of the FEIS can be used which merely attaches any changes or required findings to the DEIS. Normally, however, the DEIS is revised to become the FEIS. Section II, Alternatives, now includes a discussion identifying the preferred alternative, the reasons it was selected, and the reasons other alternatives were not selected. Revisions are made throughout the text of the DEIS where changes have been made in response to comments received. To enable the reader to recognize where such changes have occurred, the new text is often designated in some way, either by placing a line in the margin of the page or by showing new, revised text in italic. Sometimes the old text remains but is struck out to further assist the reader in recognizing the changes.

A new section is added to the end of the document. It can be titled *Comments Received on the DEIS and Responses.* It documents the public hearing and summarizes the major comments from the hearing. It also contains copies of all written comments received from agencies or the public, with written responses to all substantive comments. If the text has been revised in response to a comment, this section refers the reader to the page or section of text in the main body of the report.

Technical reports usually are not updated or revised as part of the Final EIS process. Because the FEIS has the purpose of documenting the choice of a preferred alternative, revised technical reports would no longer contribute to the decision-making process. Any revisions or refinements of technical studies are normally contained in the text of the FEIS or in the responses to comments received.

Upon completion, the FEIS is circulated to all interested agencies and to any agency or person making a substantive comment on the DEIS. Notice of availability of the FEIS is placed in local newspapers and otherwise advertised to the affected community. The circulation period for an FEIS is 30 days. Availability of the FEIS is published in the *Federal Register.*

2.4.11   Record of Decision

When an agency makes a decision on the proposed action, a Record of Decision (ROD) is prepared. The draft ROD can be forwarded to the EPA with the FEIS for filing. The final ROD cannot be filed until at least 30 days after the publication of the availability of the FEIS.

The ROD states what the decision is, identifies all alternatives considered, gives the rationale for the selected decision, and states whether all means to avoid or minimize environmental harm from the alternative selected have been adopted, and if not, why not. A monitoring and enforcement program should be included in the ROD where applicable for any mitigation.

A Mitigation Report, sometimes titled differently, can be prepared separately in conjunction with the FEIS and ROD. It should list in detail the mitigation commitments made in the FEIS and establish a monitoring and enforcement program or plan for subsequent stages of the project. This report follows the project through all subsequent stages of design and construction.

2.5 Tiered Environmental Impact Statements

Another type of EIS activity and process is the tiering of environmental documents. *Tiering* refers to completing one general EIS on an overall or very comprehensive policy or program and then using more specific EISs for site-specific or smaller actions and projects without duplicating relevant parts of the general document. The more specific EISs incorporate the broader, general document by reference and focus on the particular impacts of the proposed action which are not specifically covered in the general document.

2.6 Supplemental Environmental Impact Statement

Sometimes a proposed action or project changes after circulation of the DEIS. These may be changes in the characteristics of the project or changes in the setting. When these changes are relevant to environmental concerns and have a bearing on the decision to be made, a Supplemental DEIS (SDEIS) and Supplemental FEIS (SFEIS) are prepared and processed.

The supplemental DEIS and FEIS also are prepared when an extended period has elapsed (normally 3 years) since circulation of the DEIS with no subsequent action or preparation of an FEIS. If significant changes are thought to have occurred in either the project or other available information on the setting, then an SDEIS and SFEIS are prepared and circulated in the same way as a normal DEIS and FEIS. Although scoping is not formally required for a supplemental document, it is frequently conducted, particularly if a significant period of time has elapsed.

2.7   Reevaluation

At major points in project development, the project and its potential impacts must be reevaluated to determine if the conclusions in the DEIS and FEIS remain valid. These major points are project-specific but may include a passage of at least 3 years with no action on the proposal, the beginning of the final design phase, or the beginning of the construction phase.

A reevaluation report is prepared to document that no significant adverse impacts will occur that were not identified within the original DEIS and FEIS. The reevaluation report need not be circulated to other agencies or made available to the public, but remains within the project files of the sponsoring agency.

3

Scoping and Agency Coordination

Before any analysis begins, it is important to determine the appropriate level of study and to identify the issues and concerns that will be the focus of the study effort. This process is called *scoping*. It is one of the first steps in environmental impact analysis and involves interaction with local, state, and federal agencies and with the public. The Council on Environmental Quality has prepared an excellent guidance document for scoping (CEQ 1981b) that provides more details than are presented here.

3.1 Purpose of Scoping

Scoping is used to

- Define the proposed action
- Request cooperating agencies
- Identify what's important
- Identify what's not important
- Set time limits on studies
- Determine staff requirements of the study team
- Collect background information
- Identify required permits
- Identify other regulatory requirements
- Determine the range of alternatives

The scoping process should be specifically designed to suit the needs of the individual project or action being proposed. It can be a formal, extensive process or an informal, simple process. There are many options for the extent and format of meetings, mailings, and agency and local group contacts.

3.2   Defining the Proposed Action

Scoping refers not only to the scope of studies in the environmental impact assessment, but also to the scope of the action being considered and evaluated. As discussed in Sec. 1.9, determining the appropriate scope of a project or action requires an investigation of possible connected actions that should be included in the definition of the proposed project or action. Connected actions should be combined and analyzed in a single Environmental Assessment/Finding of No Significant Impact or Draft and Final Environmental Impact Statements.

Scoping is conducted early in the environmental analysis process. Prior to scoping, it is important for the sponsoring agency to prepare an information packet describing the proposed project or action. Scoping can not be productive until sufficient information is known about the proposed project or action to identify potentially affected parties, explain the need behind the proposed action, describe a preliminary list of alternatives and environmental issues, and present a clear description of what is being proposed that can productively let agencies and the public know on what they are being asked to comment.

An Environmental Assessment/Finding of No Significant Impact threshold of study does not absolutely require a scoping process, but most agencies find it an asset and do conduct scoping. For projects and actions requiring an Environmental Impact Statement, the scoping process is required and cannot be initiated until after publication of the Notice of Intent to Prepare an Environmental Impact Statement in the *Federal Register.* If informal scoping has been conducted prior the publication of the Notice of Intent, the Notice should clearly indicate that written comments suggesting impacts and alternatives for study will still be considered.

3.3   Identifying Issues and Concerns

Probably the most meaningful reason for scoping is to determine what issues are important, i.e., what deserves the most detailed and intense study efforts. A general overview of the proposed project or action and the surrounding area will identify the obvious issues requiring investigation. For example, a manufacturing plant is proposed adjacent to a river. The plant process will produce a waste liquid, which will be discharged into the river. A key issue for study in this project will be the effect of the plant discharge on the chemistry and aquatic life of the river.

After a preliminary overview of the proposed project and an initial identification of some of the key issues and concerns for study, the study team should further identify issues through contact with agencies and with the public.

Through early coordination letters, local, state, and federal agencies with a possible interest in the project should be contacted. The early coordination, or scoping, letter will *give* the following types of information:

- A description of the proposed project (information packet)
- Identification of the project sponsor and study team
- A preliminary list of alternatives for project design or location
- The preliminary list of identified issues and concerns
- The proposed length of time for the study.

The scoping letter should *ask for* the following types of information:

- Any special concerns of the agency
- Any required permits
- Suggestions on additional alternatives
- Relevant information and data on the surrounding area
- Future contact person within the agency

Sometimes a questionnaire can be developed to include in the scoping or early coordination letter. The questionnaire has the advantage of enabling an agency representative to quickly fill in the blanks and return the information, rather than go through the process of preparing a more formal typed-letter response. The questionnaire can request information on required permits, the suitability of the preliminary list of alternatives, any critical issues or concerns, and a future contact person for the agency.

Some states and/or agencies have established standardized forms for agency scoping. Because the same forms are used consistently, the federal, state, and local agency representatives are familiar with the questionnaires and the process. Such standardized forms are particularly useful when the contacted agency expects no particular environmental impact due to the proposed project or action within the agency's area of expertise or jurisdiction. With little effort, the agency can go on record with an appropriate response to the scoping request.

Early coordination with agencies and local organizations often leads to identification of key issues not readily obvious from the initial overview of the project and the surrounding area. For example, the proposed manufacturing plant is to be sited on a parcel of vacant land. Coordination with the local and state historic preservation agencies reveals that the land has been identified as containing important archaeological resources. Alternative sites for the plant are suggested.

Early coordination and scoping also ensure that important issues and concerns are identified at the very beginning of the study process. The scoping activities reduce the possibility of overlooking important issues deserving detailed study. A key concern identified late in the process can have a severe effect on the overall schedule and costs of the project.

For example, after the manufacturing plant has been designed and the environmental analysis studies have been completed, the Draft Environmental Impact Statement is reviewed by the local development and land-use agency. The agency notes that a school is proposed across the street. The noise, fumes, and traffic generated by the plant will be incompatible with the proposed school. The local agency strongly recommends a different site or a different design of the entire plant facility, and it threatens to litigate the project in court if its wishes are ignored.

The above example illustrates another important benefit of the scoping process. If agencies, the public, and private organizations are contacted and asked to participate early in the study through the scoping process, any late comments or concerns raised for the first time after the Environmental Assessment or Draft Environmental Impact Statement is finished then require much less serious consideration than they would have merited if raised during the scoping process.

3.4   Identifying Negligible Effects

Another purpose of the scoping process is to identify those areas of possible impact that do not necessarily apply to the project being considered. Identifying what's not so important is just as valuable as identifying major concerns, because it focuses studies on relevant issues.

The environmental impact study process has been criticized over the years for being too time-consuming and too costly. Sometimes, the study team can get caught up in details and the compilation of data not really productive to the decision at hand. Early scoping can identify areas of possible effects with minimal magnitude or probability of occurrence where limited time and effort will be consumed in gathering information, in analysis, and in the content of the documentation.

For example, the manufacturing plant is proposed on a parcel of land that is currently a totally paved parking lot with no existing vegetation. The surrounding area also is urbanized and contains buildings and pavement. The biologist member of the study team spends months gathering data on area wildlife species. The environmental document, in the affected environment section, contains detailed descriptions of every known species of wildlife in the surrounding county, including habitat requirements, feeding habits, and population levels and characteristics. The impact discussion concludes that no wildlife species are affected by the proposed project.

In the above example, the assessment spent too much time on collection of information not relevant to the possible impacts of the project. This information contributed nothing to the decision of whether to proceed with the project, and added to the total length of the environmental document. An early coordination letter to wildlife agencies would have revealed there was little potential for impact and that no detailed studies were necessary.

Environmental studies must be conducted at all times with an awareness of the study purpose in producing better decisions. Information in an environmental impact document should focus on those issues important to the decision on whether to proceed with a project; there should not be encyclopedic volumes of information that contribute nothing to the decision at hand.

Sometimes during the scoping process, the public or agencies are reluctant to say that something is not important and tend to desire to include all possible impacts in the detailed analysis. By specifically focusing a task of scoping on identifying those less significant impacts not warranting detailed study, the scoping participants are more willing to produce productive input. This identification can be assisted by requesting or referencing a specific input on how the participants feel that the suggested analysis will actually make a difference in the decision-making process.

Although particular impact categories ruled by statute, regulation, or Executive Order can be eliminated from detailed study during scoping, the environmental document should clearly contain compliance documentation with the applicable statute or regulation. Often such compliance requires a statement of why no impact is expected, with documentation, such as copies of correspondence from jurisdictional agencies.

3.5 Time and Staff

The scoping process also assists in determining the schedule for the overall study and its parts and the staff requirements. The number of areas requiring detailed analysis and an agreement on the level of detail for study methodologies are required to set realistic scheduling goals for study completion.

The disciplines to be included on the study team and the time commitment of each also can be estimated with greater accuracy after scoping. If a project is likely to have minimal natural resources impacts, the number of biologists, ecologists, and water quality experts assigned to the project will be reduced.

3.6   Gather Relevant Information

In addition to informing interested agencies and groups of the proposed project and soliciting issues and concerns, the scoping, or early coordination, letter can serve the purpose of gathering information necessary to the study. Examples may include requesting census data, land-use and comprehensive plans, and economic data; information on endangered and/or threatened species; coastal zone management plans; or flood hazard maps.

The information requested will vary by project and by agency. The most important thing to remember is that all requested information should be relevant and necessary. Do not ask for information or data that is not needed. For example, a reservoir is proposed in an undeveloped forest area at least 20 miles from the nearest town. There is no indication that there would be any secondary effects on the towns. Information on wildlife, water quality, vegetation, recreation, etc., is requested from appropriate local, state, and federal agencies and organizations. Unless there are extenuating circumstances or secondary impacts, it would not be necessary to request all the available detailed census information on population, employment, and income levels of the towns. A brief overview of information, as a basis for explaining why these types of social and economic impacts would not occur, would be sufficient.

Similarly, a project proposed in a totally urbanized city may place little emphasis on natural resources. Scoping letters to natural resources agencies should describe the project, ask if there are concerns, and ask for identification of any permits. Little, if any, additional information should be requested.

3.7   Preliminary Development of Alternatives

A preliminary list of alternatives being considered should be included in the scoping letter. Alternatives are further discussed in Chap. 4. Depending on the type of proposed project, alternatives may include alternative actions, alternative locations, or different types of facilities. The no-action alternative is always included as a viable option and for comparison with the action, or build, alternatives.

The early coordination process should specifically ask for review and comments on proposed alternatives.

3.8 The Scoping Meeting

In addition to scoping through correspondence, sometimes a scoping meeting is held to enhance the scoping process. Some agency regulations or guidelines require a formal scoping meeting; other agencies conduct scoping through correspondence. If held, formal scoping meetings are usually for agency and government representatives. Scoping meetings for the general public are often separate and are usually the first meeting on the proposed project or action in the environmental analysis phase of study.

The scoping meeting should cover the same material as previously outlined for an early coordination letter. Receiving comments from the general public on issues and concerns, or on the preliminary list of alternatives, usually requires more guidance and effort than that required to solicit similar comments from agencies or organized groups. The public meeting requires a definitive structure, through use of such methods as questionnaires or workshop groups, to productively solicit comments. Materials to solicit comments should be developed carefully to guide the public to productive input.

For example, if the meeting splits into work groups after a brief presentation, the groups should have specific work sheets to complete regarding important issues and/or concerns. If the work sheet asks, "Is pedestrian and bicycle travel an important issue?" the answer will be yes to this generalized question. But if the work sheet asks, "Are there any commonly used pedestrian or bicycle paths between City Park and Green Street that may not be officially designated or that we may not be aware of?" the answer will be very productive to subsequent studies.

A frustration in most public meetings, but probably at the scoping meeting in particular, is that the public normally wants answers that are not yet available. It is often difficult to explain that the study is just beginning and that the purpose is to focus later study efforts on relevant issues. The goal is to lead to an adequate environmental analysis, including all reasonable alternatives and mitigation measures. It may be necessary to come back to this central purpose of ensuring a thorough environmental review several times during the public scoping meeting. If attendees get into grandstanding speeches, the focus can be redirected back to the basic goal by simply asking the speaker whether he or she has any concrete suggestions for the group on issues to be covered in the Environmental Impact Statement.

The major point to remember in the scoping meeting, and all subsequent public and agency meetings, is that the goal of public input is to make it truly meaningful to the study process. As noted in the CEQ publication *Scoping Guidance* (1981b), some agencies have complained that the scoping process produces opposite results from those intended. More and more issues are raised, which makes it difficult to focus the environmental studies on relevant issues. *Scoping Guidance* correctly emphasizes that it remains the ultimate responsibility of the sponsoring agency to select the significant issues and not defer to the public. "Thus a group of participants at a public scoping meeting should not be able to 'vote' an insignificant matter into a big issue." It is extremely important, however, to not just ignore issues raised. If determined not, in fact, to be relevant issues warranting study in the Environmental Impact Statement, all matters raised during scoping should be acknowledged and explanations should be given on why these concerns were determined to be not significant.

Relevant information gained at public meetings should then be summarized and forwarded to the appropriate study team members for consideration. Too many times public meetings are held, and ineffective follow-up activities render the meetings of little value.

A scoping report or other postscoping document can be useful in summarizing major issues and concerns raised during the scoping process and giving suggestions for alternatives and mitigation measures. The document can be a useful tool for notifying agencies and the public of the decisions made on which issues and alternatives to cover in the Environmental Impact Statement. The scoping report approach is particularly useful if scoping was accomplished through correspondence or small, informal meetings. The report should list the impacts and alternatives selected for analysis and may include a more detailed written plan of study.

3.9   Public Meetings and Hearings

One of the first activities for all but minor projects is the development of a public participation plan or program. Elements of the program should include the number of meetings, timing of meetings related to study activities, format of each meeting, location of meetings, and other information to be used to distribute and obtain information, such as newsletters, hotlines, newspaper coverage, or public television coverage (call-in) programming (Fig. 3.1).

The purposes of public participation are to

- Give information
- Get information
- Establish credibility
- Resolve conflicts

A mailing list should be prepared for informing the public of future meetings and for distribution of newsletters and other information.

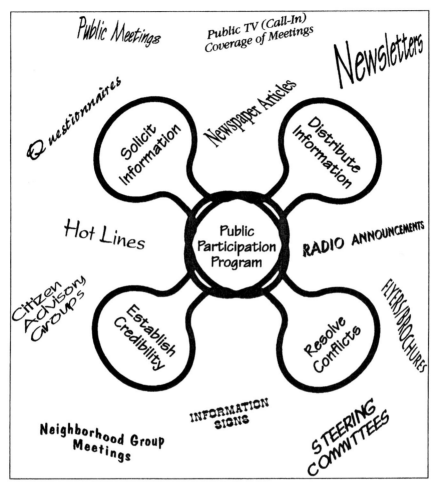

Figure 3.1 Public participation functions and tools.

Computer programs are now available that can print out names and addresses of residents by census tract or block divisions. Such information is normally available from local or county agencies. Additionally, local planning agencies can assist in identifying any organized neighborhood groups in the project area. It may be useful to hold individual meetings with such neighborhood groups or organizations, depending on the characteristics and issues of the specific proposed project or action.

At certain times in the project schedule, public meetings are particularly useful:

1. When the list of alternatives being considered has been refined

2. When the alternatives have been designed to a greater level to provide additional information on design details, such as preliminary right-of-way boundaries for a highway project

3. When preliminary results of environmental studies are available

4. When the draft environmental document becomes available for public and agency review

In many situations, the meeting held during the review period for the draft environmental document will be a hearing rather than a less formal meeting. Public hearings, however, do not need to be formally structured. Experience has shown that a good format for any public meeting or hearing is an open house. The open house has the advantage of providing a longer time for attendance; an informal, hands-on opportunity for the public to review engineering drawings, environmental results, and other information; a one-on-one interaction with members of the study team to ask specific questions; and a lack of opportunity for grandstanding or presentation of incorrect information, as may occur with a more formal type of hearing format where individual speakers are given a certain length of time and there is no discussion or feedback. An open house can be an official hearing by providing official court recorders to take statements in a separate room or area.

3.10   Continuing Public and Agency Coordination

Through public meetings, newsletters, newspaper coverage, circulation of flyers, or perhaps the formation of citizens' advisory groups or steering committees, the public should be kept well aware of the progress of studies. Often it is difficult for the general public to understand the length of time required for particular studies. It is therefore important for the study team to make periodic contact with the general public to reassure that the project studies are proceeding. In some cases, the purpose may be to explain why particular delays have occurred and the environmental studies are not proceeding according to the original schedule presented during scoping.

The public participation program is key to establishing the credibility of the study and staff with the local community. It is also the major ingredient to identification of real issues, concerns, and opportunities.

As discussed in subsequent chapters of this text, agency coordination and consultation will continue throughout the environmental impact assessment process in many particular areas of study. In some areas of potential impact, agency consultation arises in developing appropriate study methodologies, field surveys, opinion on impact analysis conclusions, and concurrence that the studies have been conducted in sufficient depth to accurately identify impacts. Specific agency coordination and consultation required by law or regulations must be documented in the draft or final environmental document as proof that compliance with the law has been successfully accomplished.

Efficient agency coordination is essential to the successful completion of the environmental analysis and to the overall timely progress of a proposed project or action. Most agencies have specialized areas of expertise and jurisdiction. Because of this differing responsibility, conflicts can occur when several differing agency resources or responsibilities are involved in the assessment of a proposed project or action. An example of this potential conflict is highlighted in the discussion of the Environmental Protection Agency's and U.S. Department of Transportation's differing responsibilities related to growth in Sec. 5.4 and in Fig. 5.1.

When agencies cannot coordinate and compromise, the environmental impact assessment process becomes deadlocked. Schedules and progress are shattered. The purpose of the process—to lead to better decisions—is defeated.

The causes of a lack of productive agency interaction on a particular project may be many. Lack of adequate staff or budget may delay timing of necessary meetings, field reviews, or comments on a proposed methodology. Often, however, the cause relates to the differing responsibilities noted above and the failure to compromise or make successful efforts to resolve conflicts. Two examples of potential difficulties are considered below.

The first difficulty sometimes experienced is the inability of the participating agency staff member to make a decision. In some cases, the staffer may not be authorized to make decisions; in other situations, the staffer may be reluctant to make decisions because of a fear of subsequent consequences and/or a lack of accountability. For example, if an agency does not make a decision to compromise, or accept a particular methodology, or agree with impact analysis results, there can be no future ramifications, such as criticism or litigation.

An example of a timely decision problem was a project located within a geographical area possibly containing endangered species. Surveys for the species needed to begin immediately, to meet project schedules and to ensure appropriate time-of-year sampling. Before any surveys for the species were conducted, written agreement on the proposed methodology, or protocol, was requested from the agency with jurisdiction. The agency staff noted that the currently accepted protocol was being revised and the new methodology would be ready in an estimated—but not certain—6 months. The staff indicated that if the current protocol were used, the survey would only have to be repeated with the new protocol at a later date, and they refused to agree to any protocol. This agency response exhibits a lack of accountability and a total disregard for productive agency coordination.

The second example sometimes occurs with relatively inexperienced staff who have been in their position for a short time. The problem can arise especially in situations where the agency has jurisdictional authority in the form of issuing a permit or signing off on an analysis when such permit or sign-off is absolutely required for the proposed project or action to proceed. The feeling of power that the staffer has over what may sometimes be a very large and important project can sometimes result in an attitude of obstinacy and unwillingness to compromise.

Fortunately, the above examples are most often the exception rather than a common occurrence. Agencies should, however, carefully review their staff and policies to ensure full and timely consultation and coordination in the spirit of the important mandate of NEPA and the environmental impact assessment process in general. Such interaction is essential to efficient and beneficial decisions in considering the environmental effects of proposed alternatives.

4

Alternatives

The development and analysis of alternatives form the very core of environmental impact assessment. An environmental impact assessment is actually a comparative analysis of alternatives. Environmental Impact Statements are often titled Draft (or Final) Environmental Impact Statement/Alternatives Analysis. The driving impetus for conducting environmental impact studies is to comparatively present the effects of proposed alternatives which, in turn, leads to better decision making.

Because of its inherent importance as the essence of impact analysis, the study of alternatives should be a thorough and systematic process. It should include input from federal, state, and local agencies and from the general public. Decisions made at every phase of analysis should be logical and documented upon a solid platform of evaluation criteria. The alternatives section of the Environmental Assessment/Finding of No Significant Impact or the Draft and Final Environmental Impact Statements is the most consequential portion of the environmental document.

Transportation projects are the predominant examples used in this chapter because of such projects' applicability to explaining the alternatives process and because the majority of Environmental Impact Statements prepared in the United States are for transportation projects. The concepts are, however, applicable to other types of projects or actions.

4.1 Purpose and Need

The foundation of alternatives development requires first establishing, in detail, the deficiencies of the status quo and the need for any action at all. For transportation projects, the need is based on the deficiencies of the existing transportation system, which may be safety, insufficient capacity to handle existing or projected traffic volumes, or perhaps an inability to meet air quality standards for a region. A National Forest Management Plan may require updating because of a regulatory requirement for timely reevaluation, a change in use demand or objectives, or an indication that the present management techniques are not producing the desired results. A new prison may be proposed because of overcrowding at existing facilities. A new low-income housing project may be based on a deficiency of supply versus existing and projected demand.

The study of purpose and need has three basic steps:

1. Define the deficiencies of the existing circumstances.

2. Determine specific needs based on the defined deficiencies.

3. Establish explicit goals and objectives to meet the needs.

The written purpose-and-need discussion will become the first section of the Environmental Assessment (EA) or Draft Environmental Impact Statement. It should be written in a clear, logical manner that methodologically leads to the adopted list of goals and objectives for the proposed project or action. Depending on the type and size of the proposed project or action, review of and concurrence with the purpose-and-need summary should be obtained from interested federal, state, and particularly local agencies and officials.

The importance of this step in the environmental impact assessment process cannot be overemphasized. The goals and objectives will form the basis for the later development of evaluation criteria by which to compare results of studies among proposed alternatives. Establishing the goals and objectives first, before any environmental studies commence, ensures a credible and legitimate alternatives evaluation which cannot be accused of being slanted or influenced by environmental study results.

For example, subsequent chapters on specific disciplinary areas of impact assessment include discussions of several laws and regulations which refer to *practicable and feasible* alternatives. Because needs and objectives are established first, and receive concurrence from appropriate agencies and officials, any alternative that does not meet the needs and objectives can legally and legitimately be considered not practicable or feasible.

4.2 Developing a Preliminary Range of Alternatives

If the purpose-and-need activity has been efficiently completed, the development of an initial range of alternatives will logically follow. The range should include consideration of all possible ways to respond to identified needs and meet established goals and objectives, including actions that may be outside the jurisdiction of the project-sponsoring agency.

For example, alternatives developed to respond to a transportation deficiency may include

- Constructing a new highway at the location of the problem

- Constructing a new highway or widening an existing route at another location that may divert traffic away from the problem area

- Widening existing highways

- Providing HOV (high-occupancy vehicle) lanes

- Providing increased bus service

- Constructing or extending commuter rail systems

- Improving traffic signal timing, adding left-turn lanes or other such measures to improve traffic flow

- Implementing inspection and maintenance programs to check vehicles for emissions

- Switching to natural gas vehicles to limit air pollutants

- Encouraging major employers to offer incentives for carpool employees

- Encouraging major employers to implement staggered work hours

- Recommending that proposed major traffic generators, such as shopping centers, major employers, or housing developments, be located in alternative geographic areas or sites

- Coordinating with local planning officials to control potential future traffic problems through rezoning or limiting permits

The above examples illustrate that the initial range of alternatives should not be limited by preconceived solutions. A full range of options that could respond to the identified needs should be created for initial consideration. The examples also illustrate the value of establishing specific goals and objectives. If, say, the stated goal is to improve traffic flow conditions, any of the alternatives may meet that goal because it is too general. Therefore, it would not be possible to comparatively evaluate the responsiveness of each alternative in meeting established needs during the alternative screening process.

Depending on the type of project being proposed, the initial set of alternatives may center on a comprehensive range of alternative sites, or locations, for a particular facility. The preliminary site analysis study will include a list of all possible sites that meet identified needs. Needs in this situation may include such items as minimum required lot size, accessibility to transportation facilities, or compatible adjacent zoning.

4.3   The Screening Process

The next step in the logical, systematic process is the screening of the initial set of alternatives to reach a preliminary list of reasonable alternatives for the initiation of scoping. For this activity, a list of preliminary evaluation criteria is developed. The list may include such items as

- Efficiency in responding to stated needs and objectives
- A maximum cost based on available funding
- Political acceptance
- Expected public controversy
- Design feasibility
- Site size or configuration requirements
- Engineering constraints
- Significant, obvious environmental effects

The environmental study will be the environmental overview discussed in Chap. 2. Sometimes referred to as a *constraints* assessment or *fatal-flaw* analysis, the environmental overview will investigate the proposed project or action characteristics and location for possible severe environmental effects. Examples may be

- Existence of an endangered-species-critical habitat
- Wetlands destruction
- Excavation required where there is an existing cemetery
- Required destruction of a National Register historic or archaeological site
- Potential severe noise impacts on hospitals, schools, museums, or libraries
- Required destruction or alteration of a significant visual amenity
- Location within a Native American religious site

- An obvious contribution of air pollutants to exceed national or state air quality standards

- An obvious discharge of wastes which will not meet water quality standards or hazardous materials regulations

- Incompatibility with zoning and land-use plans

- Alteration of a designated floodway

This list is meant not to be inclusive, but to give an example of the types of potentially severe environmental impacts that can be reasonably predicted with relatively minimal effort and within a short time.

Results of the initial screening will be a preliminary list of possible alternatives for use in the scoping process. After scoping has been completed, the received input on environmental effects or alternatives should be used to further refine the alternatives to be considered in the environmental studies and discussed in the environmental document.

The alternatives will continue to be screened and refined during the conduct of environmental analysis to avoid or minimize potential impacts. Sometimes such changes may be minor adjustments to design details, changes in particular elements of a proposed plan, or minor relocations. Continual interaction and communication among all study team members ensure that the final alternatives in the environmental document have incorporated all possible measures to reduce environmental effects.

4.4 The No-Action Alternative

In all studies, there is always the option of doing nothing. This option is referred to as the *no-action, or no-build, alternative.* It normally consists of maintenance of existing or proposed future conditions as presently planned, with no extensive capital expenditures.

Although the no-action alternative, by definition, will not respond to stated needs or meet established goals and objectives, *it cannot be dropped from the list of alternatives considered in environmental studies or from the environmental document.* The no-action alternative becomes the standard and basis by which future conditions of proposed action, or build, alternatives are compared. Because the results of the entire environmental analysis will be based on a comparison of the build alternatives with the no-build alternative, the outcome and decision-making process rely critically on the way in which the no-build alternative has been defined.

More than just a basis for comparison, however, the no-action alternative should always be considered a viable option for ultimate selection after environmental studies are completed. Such selection of the no-action alternative would be appropriate if results indicated a clear and significant imbalance in proposed project or action benefits compared with the adverse effects or costs.

The no-build alternative is not merely a definition of existing conditions. Because impacts are evaluated on the basis of future conditions, the no-build alternative must be defined in detail related to future characteristics without the proposed project or action. Planned and programmed projects not affected by the proposed project or action being evaluated should be included. Agencies will vary on the specific criteria for inclusion of other proposed projects or actions within the definition of the no-action alternative. Some agencies will include any proposed or planned project or activity. Other agencies will require that funding and other concrete evidence of implementation be in place. Still other agencies may require that only projects presently under construction be included.

Reasons for the diversity in the definition of the no-build alternative are often based on past experience or particular characteristics of agency-sponsored actions or projects. Sometimes including too much conjecture can lead to disastrous results, and sometimes not having sufficient foresight can do the same. For example, a commuter rail project proceeds based on ridership projections which are based, in turn, on projected housing and economic development. The employment and housing development never occur, and the expenditure of public funds for the transit project is seriously incommensurate with the number of actual riders on the completed system. On the other hand, a new two-lane roadway is designed based on existing undeveloped conditions in the service area. By the time the new road is open to traffic, it is already jammed with congested traffic due to construction of shopping centers, major employers, or other traffic generators in the corridor. Changes for rehabilitating the two-lane road into a four-lane road will cost twice as much as if the original design had included the development in the no-build alternative definition.

One additional similar example is an environmental analysis that compares projected air quality total pollutant levels of the proposed project with the existing conditions and concludes a negative effect. It is decided not to proceed with the transportation improvement. In 20 years, traffic under the no-build scenario will have increased threefold, and the existing transportation network will not have the capacity to handle it. Congestion will be severe, and resultant air pollutant concentrations will be twice as high as they would have been if the transportation project had proceeded. If the environmental impact analysis had correctly defined future traffic conditions of the no-build alternative, the comparison of alternatives would have indicated that the build alternative would actually result in an improvement in air quality over future no-build conditions, the no-build alternative.

The analyst should be cognizant that the no-action alternative may, in its own right, produce adverse environmental effects. Sometimes the fact is overlooked that doing nothing can cause adverse effects. In the above example the no-build alternative would produce adverse air quality impacts. Another example may be a proposed urban development project without which a particular area is projected to continually decline, with an increase in urban blight and crime.

The environmental impact assessment should always compare future conditions with the proposed project or action to future conditions without the proposed project or action. The "without project or action" conditions are the no-action alternative. It is extremely important, therefore, that the no-build alternative be defined in detail and that concurrence is received from appropriate federal, state, and local agencies prior to the beginning of environmental study activities.

4.5   Contents of Draft Environmental Impact Statement

The alternatives section of an Environmental Assessment or Draft Environmental Impact Statement should begin with a description of the alternatives selection process. Although not requiring detailed complexity, this discussion should clearly establish the systematic and logical process by which the reasonable alternatives considered in the environmental analysis were derived.

The environmental document should briefly list those alternatives already dropped from consideration and the reasons for their elimination.

The document should then describe all proposed reasonable alternatives, including the no-action alternative, in equal detail. Physical and operational characteristics should be included to a level of detail sufficient for the reader to fully understand exactly what is being proposed and the differences among the various proposed alternatives.

4.6   Reassessment and Selection of Preferred Alternative

Following the public and agency review period and the public hearing, all comments should be assembled and summarized. The considered alternatives should then be reevaluated based on comments received. In some cases, minor adjustments in design or location may respond to stated concerns. Sometimes parts of two or more alternatives may be combined to produce a more acceptable solution. At other times a downsizing of the proposal may be necessary.

Comments received during the review period of the draft environmental document also may necessitate additional design or environmental analysis to respond to specific issues raised during the review.

Based upon all considerations, a preferred alternative is then selected. The evaluation and selection process is discussed in Chap. 22. Some agencies require that a separate Preferred Alternative Report be prepared at this point. Otherwise, the reasoning behind the selection process is described in the alternatives section of the final environmental document.

If the selected preferred alternative differs greatly from any of those presented in the circulated draft environmental document, and would cause adverse effects not stated in the draft environmental document, a supplemental draft document must be prepared, must be recirculated, and must go through the public and agency review process.

4.7   Contents of Final Environmental Impact Statement

The alternatives section of the Finding of No Significant Impact or Final Environmental Impact Statement focuses on the selected preferred alternative. The selection process should be described in detail, and the reason(s) for selection of the preferred alternative should be clearly stated. Alternatives presented in the draft environmental document but not selected should be summarized, and the reasons for their dismissal from consideration should be clearly stated. The Final Environmental Impact Statement is then circulated to all agencies and individuals commenting on the draft document. Public notice of availability of the final environmental document is given, and a review period (usually 30 days) is established for receipt of any comments.

Some state requirements specifically call for formal documents to be prepared at this stage of environmental studies. For example, the California Environmental Quality Act requires certification of the Final Environmental Impact Report, Findings for each identified significant environmental effect, and Statements of Overriding Considerations. The law prohibits approval of a project unless the agency (1) has eliminated or substantially reduced all significant effects or (2) has determined that any remaining significant effects on the environment found to be unavoidable are acceptable due to overriding concerns. The Statement of Overriding Concerns must be prepared to show that each unavoidable significant environmental effect is "acceptable" based on documentation that the benefits of a proposed project outweigh the unavoidable adverse environmental effects.

4.8 Record of Decision

After the review period for the Final Environmental Impact Statement, the Record of Decision is prepared. The Record of Decision is a concise public record that includes (1) what the decision is; (2) identification of all considered alternatives, specifying the alternative which was *environmentally* preferable; (3) a discussion of preferences among alternatives and the basis for each preference; and (4) a statement on whether all practicable means to avoid or minimize environmental harm from the selected alternative have been adopted. A monitoring and enforcement program must be adopted and summarized where applicable for any mitigation.

The Record of Decision basically ends the environmental impact assessment, or NEPA, process for a particular proposed project or action. Some environmental activities often take place at later stages, however, as final design activities proceed, particularly mitigation measures. The mitigation monitoring program becomes established within the Draft and Final Environmental Impact Statement process and follows the proposed project or action through subsequent phases until completion.

5

Land Use
and Development

Almost every type of action or project can produce changes on the surrounding use of the land. Some actions and projects will have direct effects, while others may induce changes or have secondary impacts. The assessment of potential land-use impacts should be as comprehensive as the particular characteristics of the project warrant.

Land-use and development impacts are closely interrelated to neighborhood impacts and economic effects. For this reason, the discussion of land-use and development impacts is often combined with community effects, relocations, travel patterns, and economic effects under an umbrella term of *socioeconomic* impacts. This chapter and the following several chapters discuss the categories of impact sometimes grouped together as socioeconomic effects.

5.1 Defining the Study Area

The term *study area* is used often in an environmental impact document, but it can be defined in many ways. For some types of impact analysis, the delineation of the study area will be very specific as required by applicable regulations or guidelines. In other areas of potential impact, the size of the study area may differ. In all cases, the study area should reflect the full reach of possible effects within the particular impact discipline being considered.

While some analysts prefer to define one study area and then try to apply it to all impact categories, every type of impact seldom will have the same geographical extent. For example, a proposed shopping center project may have a very limited study area for the analysis of relocation impacts, or homes and land directly destroyed by the clearing of land for construction of the shopping center. For the evaluation of traffic impacts, however, the study area may extend a much greater distance, as the shopping center creates, or generates, additional traffic far beyond the actual construction site.

5.2   Existing and Planned Land Use

An initial activity is coordination with the regional metropolitan planning organization (MPO) and with local planning officials and zoning agencies. This early contact is valuable to

- Determine existing and planned land use and zoning for the area of the proposed project
- Identify any particular problems
- Identify goals for land use and economic development
- Initiate continued review and coordination throughout the project study phase

Local officials live in, and represent the residents of, the area. The town will continue long after the particular project being assessed is completed. The team assessing the project must look at every step of the study phase as if they were also residents of the area and committed for the long term.

Aerial photographs are useful in delineating study area land use, particularly in rural areas. The area should then be visited in the field to confirm aerial photography or to update information provided in planning documents.

Depending on the type of action or project, a land-use map is usually contained within the environmental document. The map should clearly distinguish between developed and undeveloped land. Categories shown on a land-use map, at a minimum, should include

- Residential
- Commercial and industrial
- Institutional and parks or recreation
- Nonurban mixed

The map could include further divisions, such as separating commercial and industrial or adding activity centers, public, vacant, etc., as categories. The decision of which categories to use will depend largely on the type of project or action being evaluated, the characteristics of the local land area, and the geographic extent of the affected study area.

Consultation with local government officials, developers, business communities, chambers of commerce, and community groups will assist in defining the land-use planning activities for the area in question. Data available from the Bureau of the Census also will provide valuable information. Many states and particularly urban areas have additional sources of information to assist in defining land-use trends, development goals and incentives, and factors influencing the local economy.

Depending on the expected magnitude of impact of the particular project or action, the environmental analysis should define development and housing trends, population growth, activity centers, proposed development projects, and zoning goals. The community's attitude on growth should be described. The result of these types of activities is the definition of the plans, policies, and goals for future land use within the affected area.

Many environmental documents include a *proposed* or *future* land-use map in the description of the affected environment. The investigator should understand that committed land-use policies, zoning, and development projects should be included in the definition of the no-build alternative. Remember that the goal is to compare future conditions in the area with, and without, the implementation of the proposed action or project.

5.3 Direct Land-Use Impacts

Some projects or actions, by their nature, have direct and obvious impacts on land use by physically destroying or clearing land and implementing a new use. Here are some examples of this kind of direct land-use revision:

- A highway project with a 300-ft right-of-way width converts whatever the existing land use was to a transportation land use within that right-of-way.

- A dam constructed to create a reservoir for water supply and recreational use directly converts the previous land use to recreational use.

- A regional park constructed on land previously used as pasture directly changes the number of acres of the park into a different use.

- A city block of low-income housing structures is razed to construct a shopping mall, directly converting that land to commercial use.

In the above types of projects, the land-use change of the property should then be evaluated for consistency with existing and proposed land-use plans and zoning ordinances for that specific piece of land

and for the land surrounding the site. Is the new use of the land compatible with surrounding uses? Or with overall plans and goals for the area or region? Sometimes a secondary impact of direct land-use changes is to pressure local officials and planning bodies to then change designations surrounding the new land use, as discussed in the next section of this chapter.

Other actions also can have a direct impact on land use by having the land-use change as one of the actual goals of the project, action, or plan being assessed. Examples include adoption of a comprehensive plan by a community, creation of national forests or parks, or designation of an area as critical habitat for an endangered species. In these types of projects, the environmental impact assessment is evaluating proposed actions that contain a direct change of land use as one of their basic characteristics.

Environmental assessments of proposed comprehensive or land-use plans, or changes in previously adopted plans or zoning ordinances, are often some of the most controversial. Environmental impact studies for plans often involve extensive public involvement through many public meetings and sometimes even organized community resistance. Controversial issues can include

- Designation of open and/or park space
- Density of permitted housing
- Single- and multiple-family housing ratios
- Owner-occupied versus rental property
- Restrictions on building heights or facades
- Locations of refuse or solid waste sites
- Location of sewage treatment plants
- Parking requirements for office and commercial buildings
- Locations for schools, churches, and other community emergency, health, or civil services

The analysis and adoption of comprehensive or land-use plans are one of the most interactive types of environmental impact assessment studies and can involve many factions of the affected community.

5.4 Secondary, or Induced, Development Effects

The analysis of possible secondary, growth-inducing effects can be an important issue in the environmental analysis of particular types of projects. Development and growth can be viewed as either positive or negative, depending largely on local characteristics of the location of the specific proposed project or action.

5.4.1 Definition

A *secondary* land-use effect is one that extends beyond, and is caused by, the direct impacts of a project. Secondary land-use and development changes can arise because the assessed project induces changes or otherwise pressures a change in use of adjacent or nearby land. The evaluation of secondary, or induced, land-use effects normally concentrates on an increase in land-use intensity, although in some cases the opposite could occur.

Starting with the opposite first, types of projects that may limit adjacent land-use development are those causing constraints on the planning and goals for that land use. Typical instances occur when land values are perceived to be reduced or no longer suitable because of the adjacent introduced project. Examples include location of a sewage treatment plant or landfill near areas originally zoned for residential development; destruction of existing homesteads or small towns to create a national wilderness area; or construction of dams or dikes that change the boundaries of the floodplain of a river and therefore severely limit the acceptable level of intensity of development.

Most impact analyses of secondary development, however, will concentrate on the induced increased density of land use. Experience has shown that whether a project will induce increased density of land development is dependent on many contributing factors. A project, in and of itself, can be compatible with planned development, but not necessarily cause it.

5.4.2 Examples of growth policy conflicts

Analysis of induced development is often associated with transportation projects, and there is often a marked difference in opinion as to whether a transportation improvement will, by itself, always induce more intensified development or growth in general. The differences most often relate to the jurisdictional and legislative mandated goals of various federal, state, and local agencies. Conflict also can occur among different regions of the country, where the priorities of issues may differ.

A good example of this difference exists in southern California, where air quality is justifiably a key issue. The agency mandated with the responsibility of protecting the nation's air quality, the Environmental Protection Agency (EPA), may take the position that any increase in capacity of a highway, construction of a new highway, or even construction of a mass transit system will automatically induce growth. The theory behind this, generally, is that by decreasing the commute time, improved transportation systems open up areas farther and farther from major employment centers as feasible housing locations.

On the other hand, the Department of Transportation (DOT) is mandated to provide safe and efficient transportation for citizens of the United States. If the existing transportation system does not meet the needs of existing and proposed economic and residential development, severe constraints can be placed on the safety of the traveling public, and existing business economies could suffer as congestion and lack of good access limit the desirability of the region for the location of business activity.

Because of criticism leveled for overestimating the ability of a transportation system to induce economic development, such as around transit stations, and factoring that economic development into project cost/benefit analysis as a benefit, the Federal Transit Administration (FTA) often discourages claiming any induced development unless absolute plans are in place.

So, on one hand, one agency (EPA) is asserting that every transportation project will induce increased development (which to that agency is a negative impact), while another agency (FTA) directs that the analyst cannot automatically claim that such economic development will occur (which to that agency is a positive impact).

Motivation for these differences in opinion can be fairly obvious upon closer examination. One of the ways the EPA promotes better air quality is by slowing, stopping, or redistributing growth. An inefficient transportation system will restrict population and economic growth, which, in turn, will result in fewer automobiles, fewer factories, and better air quality. The EPA, therefore, may oppose improved transportation systems as a means to limit growth.

The irony of such a regional EPA position is that congested highways and freeways cause the worst air quality impacts when compared to free-flow traffic conditions; and that mass transit systems are a standard transportation control measure listed in plans to *improve* the regional air quality.

The Federal Transit Administration is in the business of providing federal funds for feasible projects, based on a very detailed cost/benefit analysis. The FTA is extremely conservative in the amount of potential economic growth and the resultant multiplier into the local economy, permitted on the benefit side of the analysis. The FTA philosophy is generally that the development will occur within the region anyway and the transit stations may just redistribute this occurring development. Therefore, there would be no net gain in regional development or growth attributable to the proposed mass transit system.

The Federal Highway Administration (FHWA) often is just trying its best to meet the travel demand of an area where the existing or projected future growth renders the existing highway system inefficient and possibly unsafe without improvements.

So the question is often posed: Does the growth, which is going to happen anyway, dictate the need for transportation improvements? Or does the transportation improvement induce and cause the intensified growth?

The example above highlights two federal agencies charged with different responsibilities to serve the citizens of the United States (Fig. 5.1). Both agencies, however, and all such agencies, have an overall responsibility to consider the best solution in providing needed services and at the same time protecting the human environment. Only through agency coordination and compromise can this overall service to the people be delivered.

An indisputable fact is that growth and intensified development cannot occur without the approval of the local governing body. If no

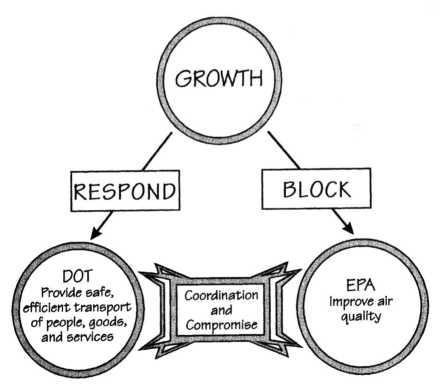

Figure 5.1 Agency differences based on primary responsibility. Agencies may be required to take differing positions, based on statutory responsibility. Recognizing that all agencies serve the good of the people, coordination and compromise are essential to prevent total gridlock.

additional building permits are issued and no changes are made to the established comprehensive plan and zoning ordinances, then no growth will occur. So, in most cases, whether "induced" development will be a secondary impact of an improved transportation system, or any other type of project for that matter, lies in the very local hands of governing bodies.

So why, then, doesn't the EPA target local governments to control growth? These local decisions are often out of the ascendancy of federal agencies such as the EPA. The alternative is to contest the proposed transportation project, which is often at least partially funded with federal funds and is therefore under the jurisdiction of the NEPA and the commitment of the Department of Transportation not to construct a transportation project that will worsen air quality in an area already not meeting the national air quality standards.

Induced growth can occur in both urban and rural areas. The most common such growth in a rural area, and one related directly to a transportation project, is the service type of land-use change often occurring around interchanges of interstate, or other limited-access, highways. If the highway passes through rural farmland, for example, the land use immediately adjacent to the interchange often is changed to support gas stations, restaurants, motels, and other types of service uses for the traveling public.

An interesting observation, on a much larger geographic scale, is the growth and development impact at the national level of the interstate highway system. Conceived and constructed with a primary original purpose of defense, the interstate highway system in the United States is one of the best in the world. Construction of this transportation system had a tremendous impact on the movement of people, goods, and services and the resultant effect on economic development and its geographic distribution.

Induced development can happen in urban, and especially suburban, areas as well, and with local streets. An example is an area of pastureland and ranches just on the border of Los Angeles and San Bernardino counties, about 35 miles east of downtown Los Angeles. Local streets were extended through the pastureland. What was previously cattle and horse grazing land quickly became housing, shopping centers, and community parks. Within 5 years, the area incorporated as the city of Chino Hills with an immediate population of 47,000 people.

Perhaps the key point in this example is that this increased density of land use was a direct result of the planning activities of local government, not the new streets. If no building permits had been issued, the new streets would be there, passing through land still supporting grazing livestock. It can only be assumed that the changed land use is consistent with the goals and objectives of local land-use policies. However, the regional freeway system connecting this location to major employment centers in Los Angeles, San Bernardino, and Orange counties is perhaps what creates the demand for housing in this geographic area in the first place.

5.4.3 Predicting induced development impacts

As is obvious from the above discussion, evaluating whether the particular project or action being assessed will cause induced, secondary development is not always an easy task.

The first step is to gather the necessary information on existing development trends, planned development projects, and especially the goals and objectives of land-use plans and policies. These existing and proposed committed projects and policies are then factored into the no-build alternative. The result is a definition of future development intensity and policy without the proposed project. The impact of the proposed project is the *difference* between these future conditions (no build) and the future conditions with implementation of the proposed project or action (build). There also may be substantive differences among the various build alternatives being considered.

Elements used to evaluate induced land-use density include

- Local zoning
- Local land-use policies and goals
- Availability and cost of vacant, developable land
- Public attitudes
- Existing constraints
- Terrain
- Surrounding land use
- Availability of appropriate infrastructure (sewage, water, power, roads, etc.)
- Supportive public policies (tax incentives, sale of vacant land at below market prices, commitment to provide required infrastructure at no cost to developer, etc.)

After the above types of factors are considered, the resultant conclusion may not be absolute. Subjective terms, such as the *degree* to which the project may induce development, may need to be used. The environmental analysis should yield the best possible prediction of environmental effects, based on available information. The conclu-

sions of the analysis of potential induced development may be that
the proposed project or action will

- Definitely cause and promote increased density of land use
- Not cause any increase in development over what would occur in
  the future without the project
- Not necessarily cause increased development, but perhaps accelerate development slated to occur anyway
- Not produce a development impact if local plans and policies stay
  the same, but indeed put into place the incentive for local planning
  bodies to change local comprehensive plans to permit higher-densi-
  ty land use

Although transportation projects have been used as examples in
this discussion, induced growth can, of course, occur with other types
of projects. Creation of a reservoir for use as recreation can result in
intensified nearby land uses to support service facilities, such as boat
marinas, motels, restaurants, and retail and commercial properties.
Building a new sewage treatment plant with increased capacity may
remove restrictions to the increased density of residential development.

Often projects causing the greatest degree of secondary growth and
development impacts are the result of private enterprise. These pri-
vately funded types of projects would most likely not be under the jurisdiction of federal regulations implementing NEPA. There may not
even be an environmental law at the state level which would apply.
Therefore, in many, many cases, the responsibility of assessing possible environmental effects will, in fact, lie with planners and officials
at the local government level.

An obvious example is the location of a major employer in a relatively small-town land use. Think of the General Motors (GM) Saturn
plant and of the casinos in Laughlin, Nevada. As each casino opened,
it immediately created thousands of jobs. Some large casinos and hotels employ 6000 to 7000 people.

In the case of the small town of Laughlin, Nevada, and the town of
Bullhead City, Arizona, directly across the river, the opening of new
casinos created an immediate problem in the job/housing ratio. There
was a tremendous housing shortage, as casinos induced population
growth. This growth was not just a redistribution of regional growth
that would have occurred anyway, for the new employment centers
drew people from all around the country. As population and housing
increased, so did the need for improved infrastructure (water supply,
sewage treatment, roads, etc.) and for increased community facilities
and services (schools, parks, fire stations, health care facilities, etc.).

5.4.4   Impacts of the impacts—assessing impacts of induced development

These types of possible impacts could be called secondary secondary impacts or impacts twice removed. If induced development is predicted, the environmental impact analysis should consider, to the extent possible, the effects of this induced development. Perhaps increased density of residential or commercial and industrial land use will, in turn, create a need for additional schools, parks, public support programs and facilities, service industries, public water or power supply, solid waste and sewage disposal capacity, improvement in local roads or intersections, or increased emergency services (fire and police) and health care facilities.

To reasonably predict the secondary secondary impact is extremely difficult and stretches the outer envelope of the crystal-ball type of prediction, rather than systematic analysis based on reliable data. The analyst should at a minimum, however, give this area of potential effect careful thought and then present it as best possible within the prepared environmental document. The document's purpose should be to present the pros and cons, as objectively as possible, or the tradeoffs of the proposed project or action. Even if an impact cannot be an exact scientific conclusion, it should be presented to induce a thought process and to ensure that even possible, qualitative issues are considered in the decision-making process. The ultimate goal of environmental impact studies and resultant documents is to produce better decisions.

5.4.5   Other secondary effects

The most common secondary effect of land-use changes is the induced growth and development discussed above. Other types of secondary impacts can occur, however, due to changes in land use or land-use plans. Many of these secondary impacts are not limited to socioeconomic effects, but can equally affect natural resources, such as water quality or wildlife habitat.

As discussed in Chap. 18 on water resources, increased covering of the earth with impervious surface, such as parking lots or large buildings, can increase the rate and pollutant loading of surface water runoff. Another example may be permitting intensified dairyland use, with extreme concentration of dairy cows on small parcels of land. Secondary effects of such use can be the increased contamination of both surface water and groundwater resources. A secondary secondary effect may then be the need to construct additional water treatment plants, with an associated secondary effect of the use of limited public funds.

The issuance of permits for increased housing or commercial development in previously undeveloped areas can have the direct effect of removing wildlife habitat. Secondary losses of additional habitat would occur over time as needed schools, commercial areas, and other secondary facilities and services are constructed.

An illustration of just a few examples of indirect, secondary effects is shown in Fig. 5.2.

5.5   Farmland Conversion

Use of land for farming requires special consideration. Agricultural usage should be indicated in maps prepared of existing and proposed land use and zoning for the project area. The actual definition of

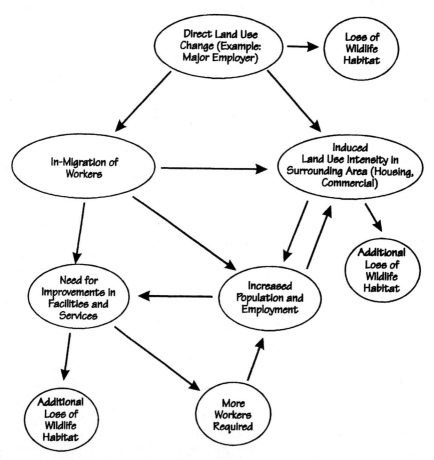

Figure 5.2   Example of interactions of secondary, cumulative effects of land-use changes.

farmland protected by federal, and sometimes state, legislative regulations does not necessarily rely on current land use, but on soils classification. Prime and unique farmland is protected by the Farmland Protection Policy Act. Farmland impact assessment is discussed in detail in Chap. 16.

5.6 Cumulative Impacts

As with all types of impact categories, the evaluation of land-use and development impacts should consider cumulative effects. A project causes a cumulative impact if, when it is considered with other proposed projects or actions in the area or region, the project produces an incremental effect to cause an overall adverse impact. The study area for possible cumulative impacts often extends beyond that of a particular proposed project or action to a larger, regional area. Examples of types of cumulative impacts include

- Cumulative demand on utilities and infrastructure
- Stress on remaining land for changes to higher density, which, in turn, may cause cumulative secondary effects, such as
  Loss of open space
  Loss of wildlife habitat
  Degradation of surface waterways or groundwater
  Air quality impacts
  Increased noise levels
- Cumulative traffic impacts due to required detours or congestion due to construction
- Long-term cumulative effects on housing or commercial service demand
- Long-term overall effects due to induced population growth on community facilities and services, such as schools, water supply, other utilities, recreational facility use, roadway improvements, etc.

The level of effort and focus on the evaluation of potential cumulative effects has increased tremendously in the past several years. Whether assessing land-use, socioeconomic, air quality, wildlife habitat, or water quality impacts, the desire by everyone is to be able to predict exactly the overall cumulative effects. Unfortunately, such a total-effect-of-all-possible-projects approach is usually not practicable or reliable; i.e., it is often a crystal-ball guessing activity with totally unreliable results.

The evaluation of cumulative impacts, however, is a real and legitimate issue. A total-proposed-all-projects approach, however, in addition to usually being impossible, does not meet the Council on Environmental Quality (CEQ) definition of a cumulative impact. A cumulative impact is the integral contribution of the proposed project or action being assessed that may, by itself, not constitute an adverse effect, but when added to the predicted effects of other projects, may be the one that contributes that "last straw" that makes the overall effect significant and adverse.

The key words, here, are the *predicted impacts* of other projects. Certainly, the environmental analysis should consider all other projects for which environmental studies have been conducted and impacts have been predicted. There may be many projects, however, within the planning stages, or even just possibly expected to happen, for which no environmental analysis has been completed. The impacts of this type of project, as sometimes with the predicted induced, or secondary, effects of the proposed project or action, most often cannot be accurately determined in any detail.

The degree of responsibility of the sponsoring agency of a particular project to then do additional environmental studies on other uncertain, possible projects proposed by others for which studies have not been concluded, remains a controversial issue.

6

Social and
Neighborhood Effects

The evaluation of social and economic impacts is an important part of every assessment. The level of detail and the amount of effort expended on this type of impact analysis vary according to the characteristics of the particular project. A project located within an urban area will most likely require the most detailed studies. Rural projects, however, also can produce significant impacts on the social and economic characteristics of an area. The difference, perhaps, is that while urban projects may require more areas and greater depth of study, rural projects, although not as diverse a study, may indeed have a more significant degree of impact.

As noted previously, the analysis of potential social and neighborhood effects is closely related to the analysis of land-use, development, and economic impacts. Neighborhood effects as discussed in this chapter refer to a more specific evaluation of identified communities within the study area.

6.1 Defining the Study Area

The definition of study area for the analysis of social and community effects requires a sufficient knowledge of the characteristics of the proposed project or action. The proposed project alternatives should be defined in sufficient detail to enable understanding of both the construction process and the long-term, or operational, characteristics.

Definition of the boundaries of the study area may also be influenced by the availability of population and employment data. Census tract data is most often used to define population characteristics. The study area boundaries are, therefore, in some cases defined by the boundaries of census tracts or block data.

6.2   Coordination with Local Agencies

Early coordination with local planning agencies and public officials is valuable to

- Obtain information and data on population characteristics
- Identify established communities and neighborhoods and the existence of organized neighborhood groups or associations
- Obtain comprehensive plans and other information listing community facilities and services
- Identify any particular problems
- Initiate continued review and coordination throughout the project study phase

Local officials and agencies are important resources in the successful evaluation of significant social and economic effects.

6.3   Data Collection

The environmental document should include a description of existing and proposed social characteristics of the study area, including demographic characteristics, identification of neighborhoods, community facilities and services, and general neighborhood character.

After a rapport is established with local planning officials, data collection can begin. Local officials will likely have information within local comprehensive plans on community facilities and services, such as fire stations, schools, and churches. The actual demographics of the area can be obtained by a review of the U.S. Bureau of the Census data. This data offers many valuable facts about an area. Be sure, however, to focus the collection and review of data on relevant issues.

Available demographic data, for instance, may be extensive. This information should all be reviewed, but with the focus of identifying special groups of population or special characteristics, such as specific concentrations of elderly persons, low income, and ethnic background. The resultant information presented in the environmental document should be limited to a brief, general discussion of the demographic characteristics, compared to averages for larger areas, perhaps the county or state. These demographic characteristics would likely include total population, income, age, ethnicity, and perhaps housing information. Population growth, historic and projected, also should be described. Any special population groups or community characteristics should be identified. If no special or unusual features are identified, the result of all this effort may be a few paragraphs and tables in the affected-environment section of the environmental document.

As the study progresses and potential impacts are revealed, more detailed data collection focuses on areas of potential impact. Such later activities may include individual meetings with community groups, interviews with school personnel, or meetings with individual fire, police, or emergency service departments.

The level of investigation and detail during data collection and in the environmental document should always be commensurate with the expected magnitude of impact of the specific action or project being evaluated. A common mistake of many analysts is to prepare an affected-environment section in the environmental document which is a huge encyclopedic volume of data. Only information relevant to the decision-making process should be included.

6.4   Neighborhood Effects

The scope of the investigation and of the analysis of potential neighborhood effects varies greatly according to the physical characteristics of the proposed action or project. The discussion in this section is not intended to imply that every item should be evaluated for every type of project. The areas of discussion are also not intended to be all-inclusive to cover every possible social or neighborhood effect. This chapter should be read with the same purpose as other chapters—to enhance and invoke the thinking process by giving examples and possible methodologies. This thinking process hopefully will lead to focused studies and investigations specific to the particular action or project being evaluated.

6.4.1   Identifying established neighborhoods

After the study area is delineated, neighborhoods within that area should be identified and described. In some locations, especially rural areas, this task can be relatively easy and straightforward. In suburban and urban areas, some of or all the following criteria may be useful in identifying neighborhood boundaries:

- An area recognized by the local population and officials by name
- An area separated from surrounding areas by physical obstructions, such as railroads, highways, or rivers

- An area of residential land use surrounded by other uses, such as commercial or industrial

- An area with a concentration of special population groups, such as elderly, low-income, or a specific ethnicity

- An area with like housing types, such as mobile homes, single-family homes, or high-density condominiums or apartments

- An area of distinct housing value, compared with surrounding areas

- An area of predominantly one type of population employment, such as professional

- An area of similar educational attainment within the population

- An area with an established community group or organization

- An area where the average length of residence in the same housing unit is more than 5 years, as opposed to areas of more-transient residents

Any special characteristics of each defined neighborhood should be identified, such as high percentage of ethnic minority population or elderly persons or a high degree of community cohesion. Community cohesion can be estimated through examination of many of the same factors listed above for identifying neighborhoods. Generally, communities with long-term residents or with established neighborhood organizations are considered highly cohesive.

The environmental document should include a map of study area neighborhoods with tables identifying special demographic characteristics, if any. This map may be combined with the land-use map, if appropriate.

6.4.2   Define future with and without conditions

As best as possible, the analyst should define exactly what physical changes will occur in each identified neighborhood. What will be different in the future if the project or action is implemented, as opposed to future conditions without the proposed project? This seems a fairly obvious task to do, but it is amazing how many environmental documents are prepared without this detailed look at and description of exactly how the physical characteristics of the neighborhoods and/or their surroundings will change.

These physical changes may include actual land clearing and destruction of part of a neighborhood, construction of new facilities within the neighborhood, changes in the physical roads into and out of the neighborhood, or changes in the visual structure of the neighborhood. These changes may also, however, be interrelated to other areas of study in the environmental impact assessment, such as predicted land-use changes around the community, air quality changes, noise effects, and increased traffic.

6.4.3 Population demographics

The proposed project or action should be evaluated for the potential to cause changes in the general population characteristics of the study area. These changes may relate to induced population growth, inmigration from outside areas, or even changes in the ratio of professionals to nonprofessionals. As discussed in the previous chapter, induced growth and development may be a direct or secondary impact. Changes may occur in the distribution of population densities. Certain neighborhoods may experience population declines, while other communities see rapid growth.

Increases in population growth and an in-migration of residents from outside areas are particularly important to consider when the proposed project would create a significant number of new jobs. Development projects typically can provide benefits to employment levels and the local economy, but can also cause adverse effects if the resultant population growth is uncontrolled, is not properly planned, or creates demand for additional services that cannot be met by private and public sectors.

6.4.4 Special population groups

The collected demographic characteristics of each neighborhood within the study area should be reviewed for evaluation of an unjust or inequitable effect on special population groups. The focus of this evaluation is usually on populations of minority or otherwise special ethnicity, elderly persons, or low-income neighborhoods. The question to be answered is whether the project or action results in an inequitable distribution of negative effects to these special population groups, as compared to negative effects on other population groups.

As with other potential project effects, the need for an analysis of this type depends on the specific characteristics of the project or affected study area. Historically, these considerations became required when proposed projects of years ago had a tendency to "target" low-income or minority communities, particularly if a substantial amount of land clearing and destruction of residential areas were required by the proposed project. Reasons for this targeting of such neighborhoods, as compared with nonminority or high-income neighborhoods, included the costs of obtaining land, lack of organized neighborhood resistance, and often political influences. Successful relocation of special population groups, however, can be much more difficult than relocation of moderate-income, mixed-ethnicity families.

If a particular project or action seems to inequitably focus negative impacts on special population groups, the analysis should continue in greater depth to determine whether such a focus constitutes an adverse effect. It is possible that certain low-income or special ethnicity areas are actually planned for redevelopment or reuse as a positive goal and objective of the overall city or community. In this case, the appearance that the impact seems to unjustly target these special population groups may, in fact, be a planned goal of the project. An example of such a project may be a proposed redevelopment project or a proposed comprehensive plan.

Some agencies have developed specific and quantitative methodologies to use in determining whether a project or action may unjustly affect special population groups. A proposed project in Texas used population density as well as other special characteristics to calculate a *potential environmental justice index* for 50-mile- and 1-mile-radius areas around the proposed project (EPA 1994). Resultant values were 20 and 8, out of 100. The document noted that the low values reflected a small population directly affected and occurred despite a high percentage of ethnic minority and economically stressed persons. The analysis concluded with a statement that "EPA has not identified any aspect of the project which appears to particularly disadvantage ethnic minorities or the poor."

6.4.5 Barriers to social interaction

Each identified community should be evaluated to determine whether the physical changes will cause direct or indirect barriers to social interaction. Does the project or action split or divide an established neighborhood? A neighborhood division can be physical or perceived. The division can be between individual residential dwellings, or a separation of the dwellings from other facilities and services that are connected through normal community travel patterns and activities. Such activities include shopping, medical visits, schools, child care facilities, and recreational facilities.

Sometimes the "division" is perceived rather than an actual physical division. An example is the other-side-of-the-tracks type of land use. Although physically not different, some city streets may be perceived as separating neighborhoods based on historic development trends or recent investments, even though there may be no physical difference between this street and other community streets.

Barriers to social interaction are frequently referred to as *community cohesion effects*. Based on the description of existing neighborhoods, those most cohesive communities would be more susceptible to a higher degree of impact.

6.4.6 Access to neighborhoods and services

Physical changes of the proposed action or project should be investigated to determine whether access within, or to and from, a neighborhood will be adversely affected. The assessment should exactly describe the changed access and should calculate the increase or decrease in travel distance or time. Access to facilities and services, such as shopping areas, medical facilities, schools, churches, and recreational facilities, should be assessed. Access also could be affected by changes in traffic volumes or patterns, even though the physical characteristics of streets, roads, or railroads do not change.

An example of such an access impact is a small town in eastern Pennsylvania that gradually over the years began to support more and more commercial establishments of a craft, antique, and artistic nature. Before long, the town became known as the place to go to find anything. Traffic volumes on weekends, in particular, increased as word spread. Tremendous pressure was eventually placed on residential streets, as homes were converted to chic art galleries and restaurants. Anyone who was anyone just *had* to visit this town. Soon the residential "holdouts" were totally incapacitated. They could not travel to community activities, go shopping, go to doctor's appointments, or even have visitors from out of town on weekends, because of the constant traffic jam and lack of parking. The physical streets, however, had not been changed.

6.4.7 Loss of residences or facilities and services (relocations)

A proposed action or project can cause a direct loss of housing, businesses, or community facilities and services. Actual relocation impacts are discussed in Chap. 8. The analysis of community effects, however, also considers the potential secondary impact of such losses on the remaining community, which can be a particularly critical impact if more than half of the neighborhood housing is destroyed. The analyst should determine whether similar facilities or services are located within the remaining community. For example, a day care center will be destroyed and relocated to another area. Does the particular affected community have other similar services? What impact might this have on child care cost or travel time?

6.4.8   Changes in property value

Property values are often a concern expressed by communities at public meetings. Although this is a social issue, it so closely overlaps with economic issues that it is discussed in the next chapter.

6.5   Community Facilities and Services

If the proposed project or action is located within a rural or sparsely populated area, the discussion of community facilities and services may be limited to a few paragraphs in the environmental document. For study areas having a more urban nature, a map should be prepared for the affected-environment section of the environmental document to show the location of community facilities and services, including, as appropriate for the characteristics of the expected magnitude of impact of the proposed action or project, the following:

- Educational facilities
- Religious facilities and cemeteries
- Fire, police, and emergency services facilities
- Health care (hospitals, nursing homes, etc.)
- Parkland and recreational land
- Civic buildings and services
- Cultural facilities (museums, libraries, theaters)

The document also should describe, as appropriate, utility services for the area, water supply, and sewage and solid waste disposal services. Note that historic and archaeological resources are dealt with separately in this text and in most environmental documents.

The assessment process begins by comparing the exact, defined before-and-after changes produced by the project or action at each identified community facility and service. Basically, the proposed project is overlaid onto the existing, or future without, conditions at each school, hospital, fire station, library, civic building, etc., to first identify and eliminate from further consideration those facilities not adversely affected.

Each potentially affected site is then evaluated in detail for the expected degree of impact. All possible impacts should be included, which will require an assembly of information from other disciplinary areas of study for the specific site being evaluated. Other kinds of information needed may include the results of air quality studies, noise impact analysis and proposed mitigation, traffic studies, or visual impact studies. Examples of the types of direct impacts that may occur and need to be quantified are given in the following paragraphs.

Actual complete or partial loss of property and relocation to a different area are the most severe type of direct impact. Such relocation may be a particularly difficult problem to mitigate because generally a replacement site would need to be found within the service area of the school, fire station, police station, church, or other type of affected service. The taking of public parkland by a transportation project is specifically prevented by a law commonly referred to as *Section 4(f)* and discussed in Chap. 10.

Changes in air quality or noise are particularly critical to schools, churches, and hospitals.

Traffic and access changes are particularly critical to response times for emergency services, such as police, fire, and ambulance services. Schools, churches, and civic and recreational facilities also may be affected by traffic impacts.

Loss of parking can be particularly critical to many types of cultural, civic, and social service facilities.

Direct requirements to relocate or revise the physical distribution infrastructure can have a specific impact on power, telephone, water supply, waste disposal, and sanitation companies.

Meetings and coordination with the affected facility or service should be conducted at this time to assist in evaluating the degree of impact. The resultant environmental document should describe these meetings and the opinions concerning affected resources.

For some reason, an environmental study is sometimes conducted totally without the involvement of those being affected. Analysts for some reason feel reluctant to coordinate with local groups, and they attempt to evaluate and make conclusions on impacts totally outside the involvement of those affected. Agencies sometimes are of the opinion that they should complete their environmental studies, present the conclusions in the appropriate environmental document, and *then* allow the required review and commentary by other agencies or by the public.

The team member assessing impacts will evaluate data and information and think to herself or himself, I wonder how the school or fire station or neighborhood will perceive my conclusions on the degree of impact? The response is, Why don't you ask them? Encourage early participation as much as possible of those potentially affected. Meet with them. Let them review and comment on your first draft of your write-up. Let them add suggestions and opinions.

The benefits reaped from this approach are numerous. The description of impact is more precise and more accurate. The early interaction reinforces the credibility of the study team to actually conduct a thorough investigation. The community feels involved in the process of the investigation, not just the results. If an agency or organization gets a chance to review and comment on draft versions of the written assessment, it feels a part of authoring the environmental document and is less likely to be surprised by what the document contains or to criticize its conclusions.

After possible direct impacts to specific community facilities or services are assessed, the next step is to consider general overall impacts to community services as a whole. These types of impacts will most likely be secondary, or indirect, effects of induced development or population growth. If the project or action will produce such an increase in growth, or perhaps a redistribution of population density, can the existing community facilities and services accommodate this new population with an acceptable level of performance?

Examples of the types of potential issues to investigate include these:

School district.   Will new school facilities be required? or a change in existing number or distribution or elementary, middle or high schools? How will the bussing routes and fleet size be affected?

Emergency services.   Will an increase in population, or perhaps even the proposed project itself, produce a need for improved or additional fire protection, ambulance, or police personnel, vehicles, facilities, or type of service vehicle?

Health and medical facilities.   Are existing facilities adequate and properly distributed to serve expected growth?

Social care facilities.   Will the increased population put stress on the operation of senior citizen centers, nursing homes, homeless shelters, welfare facilities, day care centers, meals-on-wheels types of services, transport services for handicapped or elderly, etc.?

Civic and cultural facilities.   Will a need be created for additional government facilities or services, libraries, churches, etc.?

Recreational facilities.   Will there be an overcrowding or stress on local parks and other recreational opportunities within the study area?

Public transportation.   Will there be a shift in population characteristics, or just a general growth in population, that may cause a need for improvement in public transit systems, routes, or vehicle fleet size?

Utilities.   Will an increase in population or commercial and industrial growth place demands that cannot be met on water supplies, power utilization, wastewater systems, sewage treatment facilities, and solid waste pickup and disposal?

6.6   Travel Patterns

Traffic and transportation impacts are discussed in detail in Chap. 9. As related to community and neighborhood effects, traffic and transportation impacts should consider changes to travel patterns within communities and from communities to nearby commercial or professional areas. Access to, and level of, provided public transportation to the community should be assessed, particularly in neighborhoods that, through review of census data, indicate a high dependency on public transit.

The project or action also should be evaluated in detail to determine whether the proposal would directly or indirectly encourage drivers to take alternate routes through residential neighborhoods. Particularly if the project causes congestion on particular stretches of highways or streets, motorists may be encouraged to cut through residential streets to get around the problem area.

6.7   General Lifestyle and Quality of Life

Lifestyle, or ways of life, impacts are often difficult to address and especially to quantify. Will the general character of the community change? Types of changes could include

- Subcultural variation
- Ethnicity, income, or type of work of residents
- Basic values
- Personal security
- Neighborhood stability and identity
- Leisure and cultural opportunities
- Rural versus more developed character
- Social patterns or daily activity patterns
- Aesthetic quality of the area
- General perception of individual well-being

An example is a project that creates needed jobs in a particular area or neighborhood and raises area incomes. The project would thus raise the standard of living and have beneficial impacts on the local business community, which may be encouraged to expand or upgrade the business or services. Those who are directly affected by this raised quality of life and who find economic growth to be positive would tend to consider the project impacts positive. Others in the community may be opposed to growth or to changes in general and would thus perceive the impact on their lifestyles as negative.

6.8 Public Health and Safety

Impacts to public health and safety should be based on the characteristics of the action or project being assessed. If the project would directly require employees, this section may discuss the applicable regulations that would be instituted to protect workers on the job.

Possible effects on the community as a whole should focus on aspects of the project or results of individual impact area assessments that could increase the risk of personal harm, such as air quality, increased traffic accident rates, possibility of accidental release of health-threatening pollutants into water supplies, increased chances of flooding problems, or increased crime rates.

6.9 Cumulative Impacts

The evaluation of possible cumulative social and neighborhood effects considers whether the proposed project or action causes incremental increases to an effect to such a degree as to cause an adverse effect. Although the proposed project or action may not, by itself, cause a degree of impact considered significant, will it add to impacts of other area projects to produce an overall adverse effect?

The study begins by identifying other projects or facilities proposed for the study area. As with land-use impacts, these proposed other projects are then factored into the description of future no-build conditions. The impacts of the project being assessed are added to these "background" conditions to determine whether the project will create an incremental adverse effect to such an extent as to render the total impact significant.

For example, if the proposed project or action would produce new jobs which in turn would cause immigration and growth, are there other projects being proposed in the study area which would also add jobs and population growth? Would the project being assessed increase the impact on the ability of the infrastructure to efficiently respond to the increased demands, such as water supply, fire protection, sewage treatment facilities, etc., to a point considered significantly adverse?

Another example is a landfill that will produce dust. The dust level would not be significant when the project is considered by itself. The analyst must also consider, however, whether other projects are proposed in the area that could also cause an increase in dust.

A proposed strip mine may not produce adverse visual impacts; but if it were added to other proposed projects that would produce negative visual impacts, would this strip mine's impact then become cumulatively significant?

6.10   Mitigation

Mitigation of social and neighborhood effects is often very site-specific. Techniques to avoid impacts may include providing new or revised access to communities; redesign of particular features of the proposed project or action to avoid relocations; constructing replacement open-space or recreational facilities; constructing noise control walls or security fencing; or adding parking areas.

Other mitigation measures may be intended to offset any secondary impacts of increased population growth, such as supplying accessory fire protection, security, or water and sewage capability.

6.11   Contents of the Environmental Document

The Environmental Assessment (EA) or Draft Environmental Impact Statement (DEIS) should summarize social and community effects relevant to the comparison of proposed alternatives. If large amounts of data and analysis material have been generated, they should be contained within a separate technical report supporting the environmental document.

7

Economic Factors

Closely related to social and land-use impacts, the investigation of economic factors can be a key element of the environmental impact assessment for both urban and rural areas. As with other categories of impact analysis, the amount of data gathered and presented should be commensurate with the expected magnitude of potential impact of the particular project or action being assessed. Do not allow the scope of this impact assessment to get out of hand; keep it focused on the decision at hand. If particular economic data adds nothing to the decision-making process and the proposed project or action will have no effect in that particular area, do not include it in the analysis studies or the environmental document.

Economic factors discussed within an environmental impact assessment, if applicable, may include employment and income; land values; taxes, revenues, and expenditures; economic viability of the existing business community; and proposed economic development plans and projects. Relocations also may have significant economic effects and are discussed in Chap. 8.

7.1 Employment and Income

The description in the Affected Environment section of the Environmental Impact Statement should include local employment data and information on the composition of the labor force. Normally included are the overall labor force, labor force by employment sector (agriculture, mining, manufacturing, retail, etc.), and perhaps average earnings by sector overall and per job. Employment forecasts, and more importantly unemployment forecasts, should be described. Data is available from a variety of sources, including the U.S. Census, the U.S. Bureau of Economic Analysis Regional Economic Information

System, state departments of employment or finance, and often within regional, county, or local agencies.

Impacts of the proposed project or action will be direct and indirect. Direct impacts may be those of a short-term nature during the construction stage (highways, dams, residential developments, sewage treatment plants, etc.) where perhaps the proposed project or action after construction will not provide a significant number of jobs to the local or regional labor force. Direct impacts also will occur with projects that create long-term employment (mines, landfills, major retail centers, prisons, health care facilities) for many years.

7.1.1   Direct, short-term construction effects

The direct impact on income and employment can most often be determined through application of standard coefficients, or multipliers. For example, estimated construction cost (capital expenditure) and length of construction activities can be converted to estimated number of created jobs by using standard multipliers for the construction industry for the region of the proposed project or action. The following example (adapted from Bureau of Prisons 1995) and Fig. 7.1 illustrate a method to calculate direct generated employment during construction.

Construction-generated employment:	
Project budget	$130,000,000
Labor expenditure (130 million × 70%)	$91,000,000
Average annual wage	$35,000
Total jobs over 3-year period (person-years) (91 million divided by 35,000)	$2,600
Local employment capture (person-years) (2,600 × 80%)	$2,080
Long-term (operational) employment:	
Annual payroll	$12,000,000
Payroll spent in local region (12 million × 70%)	$8,400,000
Average annual wage	$25,000
Annual created jobs (8.4 million divided by 25,000)	336

Figure 7.1   Example of calculation of direct project-generated employment.

A project budget for construction is $130 million, expended over a 3-year period, based on the construction plan and experience in the construction of similar facilities. A labor/materials expenditure ratio of 70/30 is common for this type of project. Review of U.S. Department of Labor, Wage and Hour Division publications gives prevailing wage rates for skilled and unskilled construction workers in the state or region of the proposed project or action, say, $35,000 per year. This figure includes benefits and assumes a 40-hour workweek for 48 weeks of employment annually. Based on this wage rate, the proposed project or action will result in an estimated 2600 person-years of construction employment expended over a 3-year period.

Some expertise needed for the construction project may come from laborers outside the immediate area or region, so a portion of the construction wages will not be expended locally. This percentage is sometimes estimated by reviewing journey-to-work trip distribution patterns within the region. By estimating an average percentage, say, 80 percent, of the required construction employment payroll likely to be expended within the local region, the resultant calculation is 2080 person-years of employment within the region directly supported from construction of the proposed project.

7.1.2 Purchase of materials

In addition to employee payrolls, direct construction-related and long-term operational impacts will occur through the purchase of materials, which, in turn, can create additional jobs within the industries supplying the materials. Depending on the type of materials required for the proposed project or action, estimates can often be made of the geographic location of these types of material supplier benefits. Are needed materials available within the immediate region, or will materials likely need to be transported from outside the region? Project design engineers and local contractors should be able to provide estimates of material needs, supplies, and locations. Specific goods and services required to support the construction and operation of the proposed project may also be identified through analysis of budgets for similar facilities and from examination of the purchasing and selling requirements of industries reported by the U.S. Department of Commerce, Bureau of Economic Analysis.

Estimates of the availability of needed goods and services within the local region can be determined through examination of location quotients, which measure the concentration of local economic activity in each major industrial sector. These location quotients are calculated to reflect the degree to which particular goods and services are likely to be supplied within a given region. Industry earnings data for particular regions is examined to calculate location quotients.

In the above project, where the total project cost of $130 million was divided according to the 70/30 ratio of labor to materials, the cost of materials is $39 million. As with the calculation of employment, the next step is to determine, through the location quotients, how much of this budget will be spent within the local region. It is assumed, for example, that 60 percent of the total $39 million for purchases of construction materials will be spent within the region, or $23.4 million.

7.1.3   Direct long-term effects

Direct long-term employment is usually available as part of the description of the proposed project or action. If a long-duration construction period, or a project directly creating a great number of new jobs, is proposed within an area of limited available labor force, the proposed project or action will likely cause an increase in population retention and an in-migration of employees into the study area from outside locations. The probability of this impact's occurring will directly relate to the type of created jobs (specialty or nonspecialty) and the variety of job levels offered.

Discussions with project personnel and review of historic data on employment should result in enough information to estimate (1) the split between transferred employees from outside of the area and new employee demand, (2) the type of job demand created, and (3) data from previous similar facilities on the availability of local regions to supply the required labor. Then a review of the existing labor force characteristics (gathered as part of the description of the existing environment) can be used to estimate whether the project area or region can provide the majority of the created labor needs.

As with short-term construction employment calculation, the expected number of long-term or operational, total created jobs and the percentage expected to originate in the local region can be estimated. For example (Fig. 7.1), the annual payroll is projected to be $12 million. About 30 percent of the required workers, however, are expected to transfer from other facilities, leaving 70 percent, or $8.4 million, to be spent employing persons from the local region. The type of job available will average about $25,000 in salary per year. The total number of jobs created during the operational, long-term phase of the project would thus be 336 per year.

The resultant effect on unemployment rates within the area or region can sometimes be estimated for the construction and operational phases of the proposed project or action.

The long-term, operational use of materials can be calculated in the same way as for short-term materials. Information on required operational materials and services is estimated based on budgets and previous experiences with similar facilities. The percentage of materials obtained within the local region can then be calculated based on the location quotients. For example, an annual expenditure for materials of $8 million is projected, of which 65 percent, or $5.2 million, is expected to be spent within the local region.

7.1.4   Indirect short- and long-term regional economic effects

Indirect employment and earnings are created by the ripple effect as labor force funds are respent on goods and services within the study area, inducing the production of additional goods and services by select local industries. These related, secondary economic effects also are determined by using established multipliers for the particular types of jobs within the particular study area based on estimated total expenditures. Sometimes, depending on the level of analysis, specific input-output models are developed from purchasing and selling matrix tables prepared by the U.S. Department of Commerce, Bureau of Economic Analysis. Resultant multipliers for sales output, employment, and income can be applied to the total expenditures projected for construction and operational phases of the proposed project, both for labor and for suppliers of materials.

For example, all the calculations for both short- and long-term direct employment and expenditures can be used to calculate the indirect, economic benefit through the use of multipliers. Indirect effects are calculated for wages spent in the region (after an average take-home pay after deductions is determined), for generated employment, and for materials purchases in the region. Each of these direct values is multiplied by the appropriate factors to calculate the indirect effects in sales, employment, and income.

The result is an overall estimate of economic benefit from the project, including an estimate of the millions of dollars in sales, total person-years of employment, and millions of dollars in total generated income (earnings). Often contained within the environmental document in tabular form, results are usually reported in jobs and dollars for construction and operational phases of the project. For example, the total construction and operational payroll from the project will directly support X person-years of employment; purchases of materials and services will indirectly create another X person-years of employment; and the worker household disposable income can indirectly create another X person-years of additional employment. The same is calculated for sales and income in dollars; that is, for income, an amount of direct labor income, an indirect amount based on labor expenditures, and an amount of indirect income due to purchases of materials and services.

If data is available on historic trends of distribution, estimates also can be made of how much of the regional economic benefit will remain within the study area or region of the proposed project or action and how much will likely be spent outside the immediate area. The calculated location quotients of potential supplying industries within the region also can be examined to estimate the region's expenditure capture percentage.

As discussed in Chaps. 5 and 6, an induced growth in population can have secondary adverse impacts on the community if demands for services exceed the ability of the infrastructure to manage the demand. Overall, however, a project creating employment and an in-migration of employee households will have a positive economic effect on local economies through purchases of goods and services and the multiplier effect.

7.2 Taxes, Revenues, and Expenditures

Proposed projects or actions that create jobs and induce growth will have positive impacts on the local tax base. Not only will new residents be paying taxes, but also the industry or project itself may be paying business taxes or generating revenue through other types of permits, licenses, or special assessment taxes. The economic benefits of sales revenues, jobs, and income for local construction industries, worker households, and supplier businesses will generate tax revenue to federal, state, and municipal governments.

The exception is projects that convert private property to public property which is exempt from property taxes. In some cases, the public entity will make payments in lieu of taxes and normally will pay use fees for utilities and other services. The overall impact on an individual governing unit, however, may be severe if the resultant loss in tax revenues constitutes a large percentage of the overall tax income. Such an impact is most likely to occur in rural areas.

For example, a proposed interstate highway passes through a rural small township in western Pennsylvania. The predominant revenue of the township is property taxes. The proposed highway design includes two interchanges within the boundaries of this township. The combination of right-of-way required for the highway mainline plus the additional right-of-way for two interchanges constitutes a change of private land ownership to federal. Resultant calculations reveal a potential loss in tax revenue of 20 percent of the total tax base to the township. Additionally, township roads to interchange with the new highway will need widening in the future to compensate for increased traffic volumes predicted as motorists travel from the regional area to access the new interstate. Early coordination activities with the township revealed the significance of this potential impact, and the highway design was subsequently revised to eliminate one of the interchanges.

The assessment of economic factors should include potential fiscal impacts on the municipality or county in which the proposed project or action is located. Total revenues should be described, and usually they include property taxes, sales taxes, licenses and fees, permits, business taxes, and perhaps special-use assessments, court fees, and prison fees. Project impact may be positive, direct, and indirect, as noted in the above section. A loss may occur if the proposed project or action results in conversion of private lands to public lands, or if the project would cause an extensive relocation of existing tax-base resources to locations outside the original taxing entity.

Expenditures may increase if the municipality or county needs to expand utility services or other public facilities and services related to induced growth or demanded by the proposed facility itself, such as roads, water supplies, wastewater removal and treatment, schools, or park and recreational facilities.

Expenditure of public funds for the purchase of property often is an issue raised at public meetings, particularly if national, regional, or local economies are suffering an economic depression or slowdown. Citizens quickly become aware that the amounts of public funds available to their communities are not endless and unlimited. Often, funding is not available for all important improvements identified by a particular community. Very difficult decisions must be made in selecting an appropriate expenditure of public funds on a variety of equally meaningful competing projects or improvements.

7.3   Land Values

Possible project effect on land values often is a concern raised by citizens through the public participation program. Certain types of projects, in particular, are perceived by the public as producing negative effects on the value of adjacent lands, including landfills, prisons, sewage treatment plants, highways, and sometimes commercial centers. The possibility of such an impact must be predicted as objectively as possible, using information through coordination with local officials, appraisers, real estate brokers, and historic information and studies on effects of previous similar projects. In some cases, residential and commercial land values may actually increase as a result of an increased demand for housing and commercial services.

Regardless of the results of analysis of potential impacts, for projects often perceived by the public as producing negative land value effects, an effective interaction with the local community should begin early in the study process and continue throughout subsequent study phases. As with some other types of impact categories and examples given in this text, whether the potential impact is real or perceived often makes no difference in its ultimate effect on the proposed project or action. The analyst should not be naive enough to think that, because the study team can "prove" the impact won't happen or has a minimal chance of happening, he or she can ignore or necessarily change perceived conclusions held by the public. The area of politics is one of the most powerful of all contributing factors to the success or failure of a proposed project or action.

For example, a sewage treatment plant is proposed on a parcel of land zoned for such use. Adjacent land use is also compatible, but a residential community is located a mile away. Detailed studies result in an undisputed (from scientific analysis) conclusion that the plant will not produce undesirable odors at the residential community. The community, however, has become well organized and represented. Circulated fliers (often with no scientific studies to support the conclusions) have convinced the residents that the plant will indeed produce terrible and constant odors in their community, thereby lowering property values. Local officials are pressured to deny required building permits or not be reelected. The much-needed project—proposed, in fact, to serve this very community—does not proceed.

A land value investigation is closely related to the evaluation of land-use impacts. The proposed project or action should be assessed in terms of compatibility with adjacent existing and proposed land use, as determined by local planning officials. As with the prediction of land-use impacts, other factors usually influence the potential effect on land values and can be of equal or even greater importance, such as

- Location of the proposed project relative to surrounding land uses
- Values and marketability of properties in the area prior to establishment of the facility
- Economic forecasts such as interest rates and unemployment
- Spatial distribution of availability of housing in a variety of price ranges within commuting distance
- Community and economic growth policies
- Availability of required infrastructure, such as utilities, roads, public transportation, recreational facilities, and school districts
- Special incentives for development

The potential impact on housing should be evaluated based on the projections of in-migration due to direct and secondary employment generated by the project. Availability of housing resources in a variety of type and cost ranges can be determined through coordination with local realtors, chambers of commerce, or planning agencies. An estimate can then be made of whether the proposed project or action will induce a housing demand that can be met with existing housing resources, or may require construction of additional housing. Increased demand can, of course, have a direct effect on the market status and value of both residential and commercial property in the community.

Another concern that can directly affect land values is the public's perception of personal safety and security. Particular types of projects can actually (through previous experience) or apparently induce an atmosphere around the project of creating a potential to increase crime rates, or otherwise lower the safety and security of nearby residences or businesses. An example may include a proposed prison, where the public is convinced prisoners could escape. Transit stations or large commercial centers can be thought to attract an increase in undesirable persons lurking in parking lots and traveling through adjacent neighborhoods. As with other impact categories, the analyst must assess the actual possibility of such impacts as objectively as possible, but never overlook the importance of the public's perception regardless of the conclusions of the analysis.

In contradiction to the example given previously about the sewage treatment plant, public opinion also can be overruled or ineffective in the final determination of whether a proposed project or action proceeds or dies. For example, the Three Mile Island nuclear power plant accident in 1979 irreparably damaged one of two nuclear reactors. After several years, the proposed action was to restart and put into use the undamaged remaining nuclear reactor. Public opinion was overwhelmingly opposed to the restart of the reactor. Local governments and even the state of Pennsylvania formally protested the restart. The utility company made a hardship case on the economic effects of not using this invested resource to provide electric power. The Nuclear Regulatory Commission agreed, and the undamaged reactor was put back into service, regardless of public, and official government, demands to the contrary.

7.4 Existing Business Community

The proposed project or action should be evaluated to determine the potential impact on established business communities within the potential area of project effect. Transportation projects, in particular, can affect access to businesses during the construction phase and can cause overall long-term effects to an established business community through construction of bypasses around central business districts.

Existing businesses also can be directly affected by partial land takes or other direct actions that affect parking availability. Removal of a portion of the business community through relocation (discussed in Chap. 8) may impact the economic viability of the remaining businesses in the area.

Short-term construction effects can include loss of access, public transit interruption, loss of parking, localized traffic congestion, and other actions which would discourage patrons from using the business. Retail, service, and commercial businesses are the most sensitive to temporary access impacts. Criteria that can be used to assess the potential degree of impact include

- Type of business
- Dependence on pedestrians passing by
- Size of business
- Economic stability of business

These factors should be evaluated based on the duration and characteristics of construction. Meetings can be held with the individual businesses to assist in quantifying the degree of impact. Questionnaires can be used both to gather information on the type of business, number of employees, total sales volumes, and characteristics of patrons (local, regional, pass-by, pedestrian, worker lunch-hour visits, etc.) and to assist in impact evaluation by asking specific questions of the business owner.

Results of this analysis will be an indication of those businesses which may in fact not be able to survive construction conditions for the duration of the construction period. Such establishments may be forced to go out of business. The possibilities of loss of business must be identified as objectively as possible in the environmental impact assessment document.

Some proposed projects or actions may cause a diversion of vehicular or pedestrian traffic so as to create a bypass of an established business community or district. In addition to the obvious highway bypass around a central business district, other types of projects which may have this same effect may include relocation of large employers to other locations. An example would be the relocation of state offices, employing thousands of people, from a central urban area to a suburban area. Ramifications would affect retail establishments, restaurants, mass transit patronage, parking establishments, and perhaps cultural facilities such as community colleges, museums, and libraries within the downtown area.

Providing a means for the traveling public, especially those with a regional and not local purpose, that is, through traffic, to bypass a central urban area can markedly affect service businesses, such as gas stations and restaurants, and attraction businesses, such as museums, zoos, and amusement parks. Criteria used to assess impacts can include those listed above for construction-related impacts as well as the dependency of the business on high visibility by drive-by traffic. Sometimes origin-destination surveys are undertaken to determine the percentage of regional versus local patronage.

Secondary, indirect effects of loss of businesses will include loss of employment and tax base; longer trips for patrons and employees; and redistribution of general economic activity centers within the region. Impacts of relocation are further discussed in Chap. 8.

7.5 Proposed Economic Development Plans and Projects

The proposed project or action should be assessed, in coordination with local planning officials, for compatibility or conflict with other proposed economic development projects or with economic development plans in general. Other development projects within the area of potential effect should be described, including status and expected schedule and duration of related construction activities. Potential conflicts may be direct, such as conflicts in access or land available for parking, in which case meetings should be held with the specific developers and design plans should be reviewed. More often, however, potential impacts related to other proposed development projects are cumulative.

7.6 Cumulative Impacts

As noted above, the environmental impact assessment document must consider other proposed projects in the area for examination of possible conflicts or cumulative effects. Often the considered impact area for analysis of potential cumulative impacts extends beyond the effect area of the particular project to include larger regions. Types of impacts may include

- A cumulative demand for materials and supplies during construction
- Cumulative demand on the utilities and infrastructure
- Cumulative fiscal impacts due to loss or gain in tax base or federal, state, or local municipality revenues
- Cumulative traffic impacts due to required detours or congestion due to construction

- Integral significant disruption impacts to businesses due to overlapping construction periods
- Long-term cumulative effects on housing or commercial service demand
- Long-term overall effects on local expenditures due to induced population growth on community facilities and services, such as schools, water supply, other utilities, recreational facility use, and roadway improvements

The ability to define cumulative impacts with any degree of reliability depends on the information available for other projects within the area of potential effect. Remember that a cumulative impact is the *integral contribution* of the proposed project or action being assessed that may, by itself, not constitute an adverse effect, but when added to the predicted effects of other projects, may be the one that contributes that "last straw" that makes the overall effect significant and adverse.

As noted in the discussion of cumulative effects in Chap. 5, the problem with accurate analysis of cumulative effects often is the availability of information on the predicted impacts of other projects. Assuming other proposed projects may be in very early planning stages, or even just in the "consideration" phase, economic effects of these others projects will not yet have been assessed. The analyst should, however, make the best possible estimates of cumulative effects, using information from real, committed, other projects, in either the construction phase or the final design phase, for which environmental studies have been completed.

7.7   Mitigation

The development of possible mitigation measures is most often very project-specific. In all cases, the affected community, government, agency, etc., should be consulted and informed of the preliminary results of impact analysis to then develop a set of possible mitigation techniques or actions. The list of possible measures must then be evaluated for predicted effectiveness and for feasibility.

7.7.1   Employment and income

Predictions of employee in-migration and the resultant increase in population can be offset by early planning for adequate infrastructure to meet the needs of an increased population and not cause undue expenditures for the local municipality. Often the sponsor of the proposed project or action can participate in the supply of additional housing, improved roadways, or other facilities, if it is agreed upon far enough in advance. For example, a proposed housing developer agrees to pay for, or participate in funding for, connecting streets, utility relocations or installation, public parks, and new schools, all scheduled in a timely fashion related to anticipated occupation of new housing.

Potential impacts of population in-migration and expected economic development also can be minimized, of course, by appropriate planning and zoning activities of local officials to direct distribution to desired areas.

7.7.2 Taxes, revenues, and expenditures

As discussed, public entities can sometimes make payments to local municipalities in lieu of taxes. The loss of tax base can be offset by direct design change in the proposed project, as illustrated in the example where an interstate highway interchange was eliminated.

Other types of projects may avoid or mitigate potential expenditures by local municipalities on services and facilities by including such facilities within the proposed project. Examples may include the digging of wells for the water supply, rather than putting unmanageable demands on municipal supplies, or building a wastewater treatment plant on site to process generated wastes and not adding demand to local pipelines or treatment plants which may already be at capacity.

7.7.3 Land values

Mitigation of predicted land value impacts most often is very project-specific and involves design elements of the proposed project, or surrounding parcel of land, such as landscaping to create visual barriers, staggered work hours to avoid localized traffic congestion, direction of lighting to avoid intrusion into residential areas, application of water to control dust, employ of security officers, installation of safety fencing, or other safety and security measures. Other design elements may be incorporated within the actual facility to reduce noise, avoid air quality impacts, or avoid odor impacts.

Generally, all measures to make the facility a "better neighbor" will assist in reducing any negative impacts to land values in adjacent areas. Public education programs through newspaper articles, project newsletters, or public meetings can also offset the perception of adverse land value impacts.

7.7.4 Existing business communities

Construction-related impacts can often be mitigated by limiting construction to stages, providing proper signing to inform the public of businesses still open, providing alternate parking, and scheduling the most disruptive construction activities for nonbusiness hours.

Long-term impacts are difficult to reduce in severity, but can be off-set somewhat by the use of proper signing, public awareness programs, and provision of adequate access. If an existing business can show it will lose revenue, the owner may receive compensation or be relocated to another site.

7.7.5   Proposed economic development plans and projects

The most effective method to avoid or reduce potential conflicts with proposed plans or projects is to be thoroughly aware of the specific characteristics and status of the other proposed projects. This can be accomplished through meetings with local planning officials, review of environmental impact documents prepared for the plan or other specific projects, and meetings with specific developers. Sometimes design changes can avoid conflicts. Backing up in the process to consider alternative site analysis may also be required to mitigate any impacts predicted to be severe.

7.7.6   Cumulative impacts

The mitigation of cumulative economic impacts depends on the identified type of impact, such as tax revenues and land values. Examples of mitigation measures for these impacts are given in the discussion of the specific types of impacts.

Another mitigation alternative may be possible if other proposed projects or plans can be revised to accommodate the impact of the proposed project or action being evaluated. Because these other projects normally are out of the direct control of the sponsoring agency, the possibility for this type of mitigation relies heavily on the local or regional planning officials.

Chapter

8

Relocations

If the proposed project or action will require the destruction of homes or businesses, then affected persons, families, and businesses most likely will be relocated to other sites or areas. Relocation, or displacement, impacts are probably the most significant effect of any proposed project or action on the public. The dominant question asked at public meetings for projects involving relocation is whether an individual's home or business will be taken. Although the number of required relocations may be small and determined, overall, not to constitute a significant impact compared to a project's benefits, the analyst should never forget that the impact can be devastating to those affected.

This chapter discusses the appropriate timing and level of detail of relocation studies. The contents of a general, overview study are then described, followed by the elements of a more detailed analysis. As with other environmental categories of study discussed in this text, the reader will not likely use all the examples or suggestions given. The analysis should focus on the needs of the specific proposed project or action being assessed.

8.1 Level of Detail and Timing of Studies

Assessments of potential required relocations can range from simple investigations to very detailed studies. The amount of required detailed analysis often varies considerably among various types of projects or actions, and among states, agencies, and regions. There also is often a disagreement as to the appropriate timing of accumulation of certain levels of information and analytical results.

One of the key factors influencing the appropriate type of study is the status of the proposed project or action being evaluated. If a project is in very early stages, with many alternatives still being considered, the appropriate level of analysis is an overview, a general estimate of the degree of impact. Detailed studies at this stage are precluded by

- Lack of available detail on design characteristics of the proposed project
- Consideration of a multitude of divergent alternatives, often in differing geographic locations or with differing boundaries
- High susceptibility of a project to change
- Creation of unnecessary citizen concern about individual properties that may, in fact, not ultimately be affected

Another key problem with trying to conduct relocation impact studies too early in the process is that needed information is continually changing. Therefore, studies conducted one year may be totally invalid the next year and may not reflect project area conditions at the time when the actual relocations occur.

The timing issue is in direct conflict with the environmental study process and goals. A major purpose of conducting an environmental impact analysis is to determine potential impacts before final decisions are made. The earlier that potential impacts are identified, the more this information can be used in the decision-making process. An Environmental Impact Statement is a document to present the pros and cons, or tradeoffs, of a particular proposed project or action at a stage in project development early enough to make a difference in the selection of an alternative, the refinement of a design, or the decision not to proceed with the project.

A problem that recurs often in the assessment of potential environmental impacts is that the studies conducted early, to have maximum input value, most often result in qualitative or preliminary conclusions due to lack of detailed information on the proposed project or action. These preliminary conclusions are extremely valuable in shaping the project design or the development of reasonable alternatives, but can be frustrating to the public and other interested parties because of the lack of certainty.

In an effort to reach more reliable conclusions, the need to evaluate environmental effects is sometimes used to justify proceeding with further design activities. The level of development of the proposed project or action alternatives often becomes advanced, to allow for the assessment of impacts in greater detail. A limited amount of advanced design for this purpose is permitted under NEPA and the CEQ regulations. The regulations do, however, impose limits on the degree to which a project or action may be developed prior to completion of the environmental impact assessment process. The result is that the potentially affected family or business often will not know with absolute certainty if it will be relocated until long after the completion of the environmental studies process and the selection of an alternative.

To advance the development of alternatives as much as possible for consideration of environmental effects, the sponsor of the project can expend an extreme amount of effort and funds on numerous alternatives, knowing that all but one will ultimately be a waste of effort. This fact emphasizes the importance of the process of development and screening of truly viable alternatives, as discussed in Chap. 4. The need to develop all viable alternatives to the same level of detail is now an accepted practice of environmental studies. In the years immediately following NEPA, when environmental study requirements were being developed, there was lots of information about the preferred alternative, but other alternatives were just given qualitative discussions. Rightfully, that approach was stopped because it clearly indicated that a decision had already been made prior to input on potential environmental impacts.

Another problem that often arises when detailed studies are conducted too soon is that property owners become aware that their home or business may be taken. Owners, understandably, cannot understand why there are no firm answers to their questions about relocations. They wonder how much they will get paid for their property, whether they should continue to make improvements, whether they should start to look for replacement sites, etc. Often these questions just cannot be answered in the early stages of a project design. Property owners, then, are "in limbo," sometimes for several years, and may ultimately not even be affected by the final design of the project.

The above discussion is not meant to imply that potential relocation impacts should not be studied as early as possible in the process, but rather that detailed studies may not be appropriate. Overview studies, however, are extremely critical and will add valuable information in further development stages of a proposed project or action.

8.2   Early Overview Studies

When the proposed project or action is in the very early stages and few details are known about specific design features or physical requirements, an overview type of study is appropriate as input to the decision-making process. By necessity, this early study will be somewhat simple in detail, but will have great value in identifying potential serious relocation problems. The results of the study can therefore contribute as evaluation criteria in (1) further design of the proposed project or action, (2) elimination of certain alternatives to reach a reduced set for more detailed consideration and study, or (3) a determination of overall project feasibility.

The overview study should perhaps begin with a review of aerial photography, followed by a field survey of the possibly affected area. If the proposed project or action is small and site-specific, or if several small sites are under consideration, field visits to the immediate area will probably suffice. The analyst should define the boundaries of the reconnaissance area to be larger than expected, to allow for design changes that may occur, or to actually direct the refined physical location of a project based on minimizing relocation impacts.

In addition to windshield surveys of potential properties to be relocated, data available from the U.S. Census can be used to gain an idea of overall family size, dwelling size, and price range. Sometimes local data will be available on the average number of employees per square foot of space for certain types of businesses. If data is easily attainable and reviewed, the local tax assessor also may have general information on possible displacements.

The survey should particularly investigate the presence of any special relocation problems that may constitute a "fatal flaw" in further consideration of a particular alternative, the project location, or the project as a whole. Such special uses include cemeteries, schools, churches, shelters, hospitals, nursing homes, mobile home parks, businesses requiring large sites, businesses handling materials classified as hazardous, parklands, and historic resources.

A general assessment of the availability of replacement housing or business sites can be made by looking at local newspapers or consulting local realtors, government officials, or chambers of commerce. A review of general market trends for the area will identify the likelihood of shortages of replacement business sites or housing.

As noted above, although it is a general overview by nature, this type of assessment is extremely valuable in shaping the further development of the proposed project or action, its alternatives, or various site and physical characteristics to avoid or mitigate impacts. If overview studies are conducted early enough, unwanted surprises at later stages of project processing can be avoided. The need for subsequent, extensive redesign work to accommodate late study results can thus be avoided. Particularly critical is the early identification of potentially affected parks or historic sites. Appropriate legislative requirements and processes related to historic and archaeological resources are discussed in detail in Chap. 12. Historic resources and parklands are also protected under a Department of Transportation (DOT) act, known as a Section 4(f) involvement, if the proposed project is sponsored by a federal transportation agency, as discussed in Chap. 10.

8.3   Define Displacement Area

At later stages of project design, more exact and firm physical characteristics of the proposed project or action will be known, and the area to be destroyed for construction will be that requiring displacement of existing homes and businesses. The boundaries of the displacement area should be shown on a map. The socioeconomic characteristics, such as economy, housing stock, population, and housing trends, of the general community surrounding the displacement area should be described. The description of the displacement area should include

- *Population characteristics*—elderly, minority, foreign-born, female head-of-household, mobility, transit dependency, income, non-English-speaking

- *Housing characteristics*—growth rate, single-family/multifamily unit ratio, owner or tenant occupation, size, age, price, rents

- Estimated neighborhood stability and cohesiveness, as indicated by length of time residents have occupied the same dwelling

- Major activity areas

- Existing business centers

- Land-use development and zoning

The displacement area also should be reviewed for location of community facilities and services, such as schools, health care facilities, churches, recreational facilities, and shopping areas. Results of the housing and community overviews should explain the basis of the data and the method of analysis.

8.4   Inventory of Displacements

The inventory of displaced residential units should include an estimate of the number of households, including family characteristics such as minority, ethnicity, handicapped elderly, large families, income level, and owner or tenant status. Much of this type of information can be derived from application of census tract or block data to the number of units taken. For example, census occupancy factors of persons per household can be applied to the number of parcels affected. Other pertinent information can be obtained from tax assessor maps for the area. For even more detailed studies, questionnaires can be used, and finally individual interviews and surveys of the affected families can be conducted.

The severity of displacement impacts can vary greatly with the individuals involved, and impacts are often related to demographic characteristics. If an individual is highly mobile and changes residences frequently, the impact may be only a minor inconvenience. If the community is stable and cohesive, however, and residents have lived in their homes for many years, many of those displaced may have a difficult time adjusting to new homes and neighborhoods. Special groups of residents may have strong community ties and depend upon important support networks that can be severed upon relocation. Households with school-age children may consider relocation especially disruptive if school transfers would be involved. Persons without automobile transportation also can have special relocation problems.

Criteria used to determine the structural condition of individual residential units can be derived from standards applied by the U.S. Department of Housing and Urban Development and the Bureau of the Census. Four categories are defined:

- *Excellent:* relatively new or superior housing quality

- *Sound:* only slight defects or none at all

- *Deteriorating:* no more than two intermediate defects or a single major defect but still providing safe and adequate shelter

- *Dilapidated:* several intermediate defects or a critical defect plus extensive evidence of neglect or serious damage

To establish the price range, a survey should be conducted of recent real estate sales in the project vicinity, using data from state or local market statistics. Recent sales in the vicinity can then be listed by price range, floor areas, unit values per square foot, mean sales price, median sales price, and size (number of bedrooms). The distribution of sale prices is helpful, such as $40,000 to $59,900, 10 percent; $60,000 to $79,900, 45 percent; $80,000 to $99,900, 30 percent; and over $100,000, 13 percent. Values obtained by this approach should be supplemented with opinions expressed by local real estate agents. The number of renter-occupied residences can be determined by assessor rolls where a residential unit has a different address for the owner.

The loss of a substantial number of houses and apartments affordable to people with low and moderate income may have an effect on the community's affordable housing stock. A low inventory of affordable housing can increase housing prices and disproportionately affect low- and moderate-income groups. A methodology should be established and described for calculating affordability and for defining low- and moderate-income households. The number of affordable housing units subject to removal should be compared with the total number of affordable housing units in the community.

The Environmental Impact Statement should contain summary tables of relocations indicating the number of residential units, divided perhaps into single-family (SF) and multifamily (MF), and the estimated number of households, families, or persons. Summary tables on housing may include housing type, price range, rental rates, vacancy rates, SF/MF ratio, size, age, and condition for each structure subject to full acquisition under each alternative.

For projects requiring detailed studies, a separate technical report should be prepared covering methodologies, data tables, detailed analysis, and conclusions. For residential units, a typical table may contain the following headings for each alternative: housing type; address; census tract; rental (checked or not); number of units; interior structure (5 rooms/3 bedrooms/1 bath); year built; land size; improvement size; and condition (excellent, sound, deteriorated, dilapidated).

Results of the inventory of displaced residential units should include a summary of special cases, such as elderly, handicapped, transit-dependent, low-income, required proximity to health care facilities, very large families, non-English-speaking, mobile homes, or other factors which may require special consideration in relocation. Local school district officials should be contacted to determine whether there would be any potential impacts associated with the reduced attendance related to projects requiring extensive relocations.

8.5 Relocation Resource Area

The next step in a relocation impact study is to identify a nearby area suitable for moving the displaced families. Criteria used to determine the adequacy of replacement housing and the suitability of the relocation resource area may include

- Proximity to displacement area
- Comparable housing types, age, condition, and prices
- Comparable rental prices and vacancy rates
- Proximity to major activity centers
- Access to transit routes
- Location within same school district
- Community demographic characteristics (income, ethnicity, minority, age, educational attainment)

The identified relocation resource area should be defined with boundaries and a map in the environmental document. The selected area should be compared to the displacement neighborhood with relation to schools, churches, shopping, public transportation, community and recreational facilities, hospital and medical facilities, and employment areas.

Realtor associations and multilisting services can be of use in defining the boundaries and the resources within the identified area. The environmental document should include a table of the total number of single-family dwellings currently available for sale in the resource area by price range, size, and condition. The available housing resources should then be compared with the identified need of the relocatees.

Because, unlike single-family residences, apartment units for rent do not have an organization that functions as a clearinghouse, as does the multiple-listing service, units for rent can be determined from the use of the local Sunday newspapers. Choose specific locations to define the resource area. As a supplement, the local apartment guide can be surveyed for rental units. Results should be listed in a table giving the price range and size. Available single-family residences for rent also should be listed by price and size, with information derived from local newspapers.

Because of the complex nature of mobile homes, a complete, separate analysis should be made. The description of units affected should include park type (rental rates, adult, retired, family, etc.) and home type (age, condition, length, width, number of bedrooms, etc.). Analysis for relocation potential will include a list of parks in the vicinity, park locations, space available, rental rates, tenant requirements, and park requirements. If new park development will be necessary for replacement sites, indicate the general attitude of the community toward mobile home parks; vacant land available for development; vacant land zoned for mobile home park development; and if none is available, the attitude of the local planning commission toward rezoning for this use.

When the relocation process is actually at hand, personal interviews should be conducted with those to be moved, in order to gather information related to

- Preferences concerning the area of relocation
- Number of people to be displaced and distribution of adults and children
- Location of schools and employment
- Special arrangements needed to accommodate any handicapped member of the family
- Financial ability to relocate into a comparable replacement dwelling which will house all members of the family decently

Residential replacement dwellings should be in equal or better neighborhoods at rents or prices within the financial means of the individuals and families displaced, and be reasonably accessible to their places of work.

8.6 Nonresidential Relocations

Nonresidential relocations may include commercial and industrial or institutional and public service. Commercial uses would include retail and wholesale, industrial and warehouse, service-oriented, and office. Institutional relocations include churches, parks, schools, hospitals, government structures, charity organizations, utilities, and other public buildings.

Commercial and industrial relocations should be described, including the number taken, type of business, type of occupancy (owner or tenant), and number of employees. In addition to field surveys of the affected area, the number of parcels, amount of land area, and square footage of improvement can be obtained by a review of the tax assessor's information. The number of employees can often be estimated by applying standard factors of employees per square foot by type of business from the study area or comparable cities.

If a detailed level of effort is warranted, questionnaires can be circulated to businesses in the area. Information requested will include the type of business, size, employee method and distance of travel to work, patron service area (local versus regional), and particular problems with relocation to an alternate site. The final level of detail will include personal interviews with affected businesses.

As with residential displacements, detailed information, such as lists of individual businesses by name and address, should be contained within the technical report prepared in support of the Environmental Impact Statement. The Environmental Impact Statement should include a summary table listing the number of businesses and number of employees affected.

Institutional relocations, such as of government buildings, libraries, parks, homeless shelters, hospitals, churches, or schools, will require more detailed descriptions, including the function of the facility, service area, number of employees, number of persons served or accommodated by the facility, and any special location requirements. Sometimes these types of facilities have special requirements for proximity to other similar facilities, for ease of access for local neighborhood pedestrian patrons, or for maintaining a location within the same service area, which may be quite small.

Because of the special sensitivity of institutional and public service relocations, hopefully the proposed project or action will have been designed to avoid the need to relocate this type of community resource. Potential institutional and public service relocations are a good example of the impacts that should be identified early in the environmental overview studies to shape the development of the proposed project or action to minimize impact. If such an impact is unavoidable, however, extensive meetings with affected facilities are normally required to assess the degree of impact and relocation requirements.

When the inventory of nonresidential relocations has been completed, it should be reviewed to identify potential special cases or specific relocation difficulties. Relocations may require special consideration because of (1) large required site size; (2) use of materials classified as hazardous (replacement sites would need to be in areas permitting such activities); (3) a requirement of visual exposure to nearby roadways, for service industries, such as restaurants, motels, and gas stations; or (4) a need to remain within a small service area.

As with residential displacements, an analysis should be made of the availability of suitable replacement sites for relocation of commercial, industrial, institutional, and public service facilities. Contact with local chambers of commerce, realtors, and government officials can be used to identify the availability of vacant properties in appropriate ranges of cost and zoning. Criteria to indicate suitability of a replacement site will likely be much more individually specific to the needs of the relocatees.

Based on the results of the analysis of characteristics of nonresidential displacements and the availability of suitable relocation sites, an evaluation should be made to predict whether any of the affected facilities is expected to cease operation due to displacement. Business size, operational characteristics, and financial resources should be evaluated for indications of closure or substantial loss of business due to relocation.

8.7 Cumulative Impacts

Any other proposed projects in the vicinity should be identified and examined to determine whether relocations also will be required. Resultant competing displacement needs should be quantified as much as possible, and feasible measures to minimize impact should be investigated in coordination with sponsors of the other projects and with local government officials and agencies.

8.8 Mitigation

Assistance should be provided for all persons and businesses required to relocate. The fair market value is offered for the purchase of existing property. Contacts with local governments, organizations, and groups can assist in identifying measures to reduce impacts. As described in the discussion below, monetary assistance is usually provided. Regardless of the mitigation measures taken, relocation of families and businesses will most likely remain a significant, unavoidable adverse impact.

Special measures are required where the existing housing inventory (1) is insufficient, (2) does not meet standards, or (3) is not within the financial capability of the relocatees. Similarly, detailed studies will be needed for business relocations if relocation sites are not available or do not have appropriate characteristics. Relocation of institutional and public service facilities will probably also require specific mitigation measures to ensure the minimum possible impact.

The Uniform Relocation and Real Property Acquisition Policies Act of 1970, as amended, requires that relocation assistance be provided to any person, business, or farm operation displaced because of the acquisition of real property by a public entity for public use. Compliance with this federal act is required by any public agency where federal funds are to be used in the acquisition or construction of the proposed project. In all cases, relocation resources are available to all residents and businesses without discrimination. Many state laws also have provisions that require the sponsor of the proposed project or action to provide payments and services to persons displaced by a public project.

Payments can include

- Moving costs

    Purchase supplement (for increased costs of replacement housing) in the form of price differential payment (when the cost to buy a replacement dwelling is more than the amount being paid for the displacement dwelling)

    Payments for certain nonrecurring costs incidental to the purchase of a replacement property

    An interest differential payment if the interest rate for the loan on the replacement dwelling is higher than that of the loan on the displacement dwelling; then gives maximum of the total of these three supplemental payments, and if total entitlement is in excess, then last-resort housing program is used

- Rent supplement, as usually a lump sum with a maximum amount; if over that amount, housing of last resort applies.

- Downpayment option (with maximum)

- Housing of Last Resort Program (49 CFR Section 25), used when displacees cannot be relocated because of the lack of available comparable replacement housing, or when their anticipated replacement housing payments exceed the limits of the standard relocation procedures

Moving costs for businesses can include actual moving expenses and payments *in lieu of* actual moving expenses. Payments can include actual, reasonable moving and related expenses; actual direct losses of tangible personal property; and actual reasonable expenses for searching for a replacement site. Payments in lieu of moving will apply if the business elects to receive a payment equal to the average annual net earnings of the business (in cases for businesses expected to suffer a substantial loss of existing patronage as a result of the displacement). Limits are set on all these payments.

8.9   The Relocation Plan

When appropriate in project development, a Relocation Plan should be developed. Begin by stating the basic assumptions, which could invalidate all or part of the study if changed, such as the date of project completion, that the design will remain essentially unchanged, critical recommendations in the Plan are implemented, and all approvals are obtained as scheduled. The Plan should do the following:

- Relate housing needs to availability.
- Describe those classes of relocation where no special effort will be required.
- Describe mobile homes separately.
- Describe problems where the normal market may not have enough decent, safe, and sanitary housing to absorb the demand within the time span allowed for relocation.
- Prepare a time schedule of the best estimate of time required.
- If there's a shortage of housing, look at previous project studies to try to generalize the percentage of displacees by housing class who tend to leave the area completely.
- Survey to find out if there is a housing area with a large number of ineligible occupants, such as student housing and motel and hotel occupants.
- Note which parcels should be appraised and acquired first.
- Indicate which parcels for which extra time will be a solution to special problems, such as very large houses, rest homes, old hotels with permanent residents, housing for elderly people, and mobile home parks.

If comparable decent, safe, and sanitary replacement housing is not available and cannot otherwise be made available, the public sponsor must use project funds for the construction of, or otherwise make available, necessary replacement housing. This may be in the form of payment beyond the basic program. Should construction of new replacement housing be required, the Relocation Plan must specify, in detail, the location, costs, and schedule of construction and relocation activities.

For example, a major freeway in Los Angeles required the substantial displacement of low- and moderate-income families and individuals, and comparable replacement housing was not available in nearby communities. The first construction activity of this major project was to build numerous high-rise condominium and apartment buildings for the relocation of the displacees.

9

Traffic
and Transportation

Certain types of proposed projects or actions can have impacts on travel patterns, traffic volumes and flow, pedestrian access, and bicycle travel. The evaluation of traffic and transportation impacts is closely interrelated to the assessment of land-use, social, economic, air quality, and noise effects. This chapter presents basic introductory information on traffic analysis and potential impacts that should be considered. It is not a substitute for a qualified traffic engineer for projects that so warrant.

9.1 Applicable Projects or Actions

These are examples of types of proposed projects or actions where traffic impacts may become a key issue in the environmental impact assessment:

- Land-use or comprehensive plans
- Proposed highway or transit improvements
- Projects that attract large volumes of traffic, such as shopping centers, amusement parks, schools, convention centers, parking structures, or municipal buildings
- Major event venues or employment centers
- Housing developments
- Changes in bus or parking rates in major urban areas
- Individual projects that may block, or render unsafe, pedestrian and bicycle travel or access for the handicapped

The level of analysis of traffic and transportation impacts will depend on the particular features of the project or action being assessed. As with other types of impacts, the level of effort should be commensurate with the expected magnitude of impact and with the results of the scoping process.

9.2   Traffic Analysis

Vehicular traffic on streets and highways can be assessed by using various standard traffic analysis procedures. Most projects expected to produce traffic impacts will, at the least, require a description of existing, and perhaps historic, traffic volumes and flow characteristics and a prediction of how those volumes and flow characteristics will change in the future both with and without implementation of the proposed project or action. In other words, the analysis should discuss, in equal level of detail, the future traffic characteristics of all proposed alternatives including the no-build alternative, which will be used as the baseline for comparison with the proposed build alternatives.

9.2.1   Volumes and levels of service

Traffic volume data is normally available from the state department of transportation, local planning agencies, or regional metropolitan planning organizations. Raw data is gathered by actually counting traffic volumes throughout the hours of a day at a particular point of a street or highway. Traffic volumes are normally reported as average daily traffic (ADT) and morning and evening peak-hour traffic.

Certain physical characteristics of streets and highways dictate a calculated traffic capacity for that particular facility. Examples of the features entering into the capacity analysis include the lane width, number of lanes, shoulder width, grade (slopes of hills), radii of curves, type of access permitted, and distance between ramps or traffic signals. The type of access permitted can be divided into several categories. There are three basic types of access:

1. *Free access*—at-grade intersections, adjacent joining driveways, and left or right turns possible

2. *Controlled access*—streets or highways with medians that only permit crossing at designated places, and only right turns onto or off of the street permitted except in designated areas

3. *Limited access*—the freeway, expressway, or turnpike facility where crossings of other highways and streets are grade-separated and access is limited to free-flow interchanges

An *at-grade intersection* is the typical stop sign or signal light on ground level. A *grade-separated crossing* occurs where the crossing street is carried over or under the expressway via a structure (bridge). A *free-flow interchange* provides ramps onto and off of a freeway or expressway without requiring the vehicle to stop at the freeway or expressway. Common interchange designs include the diamond and cloverleaf, but there are many varieties, each with specific advantages and disadvantages regarding traffic flow and capacity.

Capacity can be determined for the mainline of the expressway or street and for intersections and interchanges. Capacity analyses for intersections and interchanges are more complicated because characteristics such as the number of left-turn lanes, timing of signal red and green cycles, and timing of nearby signals must be factored into the analysis.

Level of service (LOS) is a qualitative measure to describe the flow or operational characteristics of traffic, as perceived by the level of congestion or delay experienced by the motorist. The level of service is a result of the transportation facility's capacity, or ability, to accommodate the volume of traffic on the facility. Also the LOS can be affected by the characteristics of the traffic itself, such as the percentage of trucks in the total traffic. Levels of service are as follows:

LOS A represents a free flow of traffic. Individual users are unaffected by others and have the freedom to select desired speeds and to maneuver within the traffic stream.

LOS B is in the range of stable flow, but the presence of other users begins to be noticeable. Freedom of speed is unaffected, but there is a slight decline in freedom to maneuver.

LOS C is in the range of stable flow, but marks the beginning of the range where individual users become significantly affected by interactions with others, hindering selection of speed and maneuverability. The general level of comfort and convenience declines noticeably at this level.

LOS D represents high-density but stable flow. Speed and freedom to maneuver are severely restricted, and the driver experiences a generally poor level of comfort and convenience. Small increases in traffic volumes will generally cause operational problems at this level.

LOS E represents operating conditions at or near capacity. Speeds are reduced to a low but relatively uniform level, and maneuvering is extremely difficult. Comfort and convenience levels are extremely poor, and driver frustration is generally high. Operations at this level are usually unstable because small volume increases or minor fluctuations will cause breakdowns.

LOS F represents forced or breakdown flows, and it exists when the amount of traffic approaching a point exceeds the amount which can traverse the point. Queues form behind such locations, and the extremely unstable operations are characterized by stop-and-go waves.

Level of service is governed by traffic density, measured in passenger cars per mile per lane. The density is converted to passenger cars per hour per lane, using the average speed of the traffic stream. Generally, levels of service A, B, and C are considered good operating conditions with only minor delays. LOS D represents fair to below-average operating conditions, but is sometimes acceptable in urban areas. Levels of service E and F represent extremely congested conditions.

Level of service for traffic analysis of signalized intersections is defined in terms of delay. LOS criteria are stated in terms of the average stopped delay per vehicle for a 15-minute analysis period.

9.2.2 Forecasts

Future traffic volumes can be predicted in a variety of ways. Regardless of the method used, it is extremely important to the environmental analyst that the exact assumptions and methodologies be documented. There should be a review and an agreement by local and state traffic specialists that the applied approach is acceptable and reliable. Results of traffic studies are subsequently used in the analysis of land-use, neighborhood, economic, air quality, and noise impacts. If there is a justified question concerning the development of forecasted future traffic volumes, conclusions in these other areas of impact analysis also are in doubt. Methods for forecasting future traffic volumes have become significant issues of controversy in many, many projects.

A relatively simple method sometimes used to predict traffic volumes involves reviewing historic data on traffic growth rates for a particular transportation facility or area and then predicting the future growth rates. The predicted future growth rates may depend on predicted employment or population growth contained within a regional or local comprehensive plan. The future growth rates are then applied to existing traffic volumes to arrive at future volumes for the target year of analysis.

More detailed methods can include the systematic division of an area into sectors or zones and then application of population and employment growth predictions to each sector. Travel origin and destination studies can be done for existing traffic through questionnaires and surveys, and future conditions can be estimated. Predictions are then made of the number of trips originating in a particular zone and traveling to another zone. Computer models are used for the complicated data input required.

For individual site analysis, estimates of potential generated traffic can be made by gathering information on the future number of employees, expected number of shoppers at retail centers, number of potential attendees at major sporting events, etc., for use in predicting total traffic volumes in future years.

For the evaluation of comprehensive and land-use plans, the included permitted density and dispersion of various land uses will determine generated traffic and should yield estimates of rates of growth.

9.2.3 Results

Predicted future traffic volumes are applied to the future transportation network, which may or may not be changed over existing street and highway characteristics, depending on the type of project being assessed. Normally future traffic volumes are given in average daily traffic (ADT) and AM and PM peak hourly volumes for comparison with existing volumes. For each proposed alternative, including the no-build alternative, future volumes should be presented in either tabular or graphical form. Often a line drawing of the street network is used to represent traffic data, with volumes shown at particular locations.

By applying future volumes to the transportation network, the level of service can be calculated. Traffic impacts of each proposed build alternative can be compared to those of the no-build alternative. The difference will be the impact of the individual project or action.

For example, a particular arterial street operates at LOS C. Future traffic volumes are expected to grow, and with the future year (say, 10 years from the present) no-build alternative, the street is projected to operate at LOS D. The proposed project being assessed is a major employer, with thousands of new jobs located on this particular stretch of street. When the projected employee traffic is added to the total forecasted peak-hour traffic volumes, the street will operate at LOS E, indicating a severe traffic congestion impact caused by the proposed project.

Caution is required in interpreting the above example results as generated traffic. The actual impact is generated traffic on that particular piece of street, leading to increased congestion during peak hours. It could be described as a change in traffic patterns. Or perhaps a particular project will "generate" additional traffic in a particular town. At the regional level, however, there is debate often over whether a major employer, or a roadway improvement, or other types of traffic impact projects actually generate a net increase in traffic, or just redistribute existing traffic into different patterns throughout the region.

In the above example, for the additional traffic on the street to actually be *generated* new traffic would require that (1) all the employees were previously unemployed and made no trips to work or (2) all employees moved into the region as a result of employment opportunities or (3) the employer would not locate somewhere else within the region, if not at this site. It is easy to see why this issue can be in question and is often not easy to answer.

Projects with geographically extensive or long-duration construction periods may also create adverse traffic impacts. In cases where an impact is expected, a maintenance of traffic plan should be prepared by the project designers. The maintenance of traffic plan will describe staged construction activities, detour routes, signing, and other measures to lessen the impact on traffic during construction.

For projects requiring construction-related or permanent truck or heavy equipment access, haul routes should be identified in advance. Mitigation measures, such as dust control, air quality control, or noise control through limiting hours of operation, should be identified and assessed for successful mitigation of impacts to acceptable levels.

9.3 Secondary and Cumulative Impacts

As with many other types of impacts, traffic congestion impacts can cause secondary effects, such as the following:

- Increased noise and air pollution
- Adverse visual effects
- Delays in emergency vehicle services
- Loss of patronage to restaurants or retail establishments due to inconvenience of access
- Increase in motorist accidents and decrease in pedestrian safety
- Reduction of ability of an area to keep major businesses or to attract new businesses
- Changes in travel patterns of through traffic into neighborhood streets as motorists attempt to avoid congested portions of major streets
- Inconsistency with goals and objectives of local land-use or comprehensive plans

It is important to review the results of traffic studies carefully to identify all possible secondary impacts.

Secondary impacts also can occur during traffic changes caused by construction activities. For example, a tunnel is being dug for an underground mass transit system. Material taken from the tunnel (sometimes called *muck*) must be hauled away on trucks. A school and a nursing home are located near the muck out site. The major urban area and adjacent intersections already operate at poor levels of service and are classified as nonattainment of specific air quality standards. When construction truck traffic is added to normal traffic volumes, increases in noise and in air pollutants are identified as secondary adverse impacts. Mitigation may be to limit the number and timing of trucks into and out of the muck out site.

Cumulative impacts should be already considered in the forecasts of future-year traffic volumes. Hopefully, research on other projects in the construction or planning stages will be identified and factored into future-year projections. It is a good idea, however, to cross-check particular projects or particular locations within the impact area to ensure that cumulative traffic effects have, in fact, been considered.

9.4 Mass Transportation Systems

Proposed new mass transit systems or bus service, or changes in existing systems, can have effects on vehicular traffic patterns on roadways. Traffic impacts may occur at major transit stations or at park-and-ride lots. If trips are transferred from individual vehicles on the roadway network to trains or buses, however, the overall effect may be to reduce traffic congestion and move more people and more goods more efficiently. Proposed projects should be evaluated for design features to encourage mass transit use, such as location near transit stations, or special incentives for employees who carpool.

Changes in parking policies or rates, bus schedules or fares, and train schedules or fares also can cause transportation-related impacts. Particularly in urban areas, a portion of the population will be transit-dependent. Raising fares can sometimes unfairly impact special social population groups, such as the elderly, low-income, or handicapped persons. Changes in schedules and fares could cause adverse accessibility effects for transit-dependent persons to travel to jobs, medical facilities, or shopping and visiting trips.

9.5 Pedestrian and Bicycle Travel

All projects should be assessed for possible adverse impacts on pedestrian and bicycle access and safety. Local and regional land-use and comprehensive plans should include information on existing designated pedestrian trails or walkways and on bicycle routes. A review of the potential area of impact of the proposed project or action can be conducted to identify major routes for nonmotorized traffic or major indications of use by pedestrians and bicyclists. The proposed project or action should be reviewed to determine whether the physical characteristics will sever existing routes or physically divide neighborhood pedestrian and bicycle movements. Linear projects, in particular, may cause restrictions on pedestrian and bicycle traffic and divide neighborhoods. Examples include wide limited-access or controlled-access freeways and streets, major drainage channels, and railroads. Nonlinear, site projects, however, also may create an obstacle to nonmotorized traffic access, particular if the site is extremely large. Examples include prisons, military bases or depots, and large industrial and manufacturing sites protected by fencing.

If the proposed project is a federally funded highway, specific legislation requires identification of severance of any existing major route for nonmotorized traffic. The selected preferred alternative or action (described in the Final Environmental Impact Statement) must provide a reasonable alternative route or demonstrate that such a route exists [23 U.S.C. 109(n)].

9.6 Mitigation

Numerous mitigation measures can be employed to offset traffic and transportation impacts. The range of possible mitigation techniques can extend from a regional level to very specific design characteristics of a particular proposed project or action.

9.6.1 Traffic congestion

A direct means to reduce traffic congestion is to increase capacity on the highways or streets operating at poor levels of service, such as by adding lanes to the roadway. Ramp meters can be installed to control the timing of the flow onto limited-access highways. Synchronization of signals can improve flow on arterial streets, and intersection operation can be improved with specific traffic control measures, such as left-turn lanes or signal cycle timing.

Traffic congestion also can be reduced by changing the characteristics of the traffic, as opposed to the highway or street. Such measures may include staggered work hours at major employers, incentives for carpools or use of mass transit, or provision of high-occupancy vehicle (HOV) lanes on ramps and lanes of major expressways (for use by only cars with two or three passengers and/or buses). Message systems that warn motorists of congested areas to avoid and a program for rapid emergency service for broken-down vehicles also assist in mitigating congestion on major routes. Transportation management techniques can be applied at a local level or within a large region.

9.6.2 Construction impacts

A detailed maintenance of traffic plan will offset many of the impacts of construction activities on the movement of traffic. Staged construction activities will provide continued operation of as many traffic lanes as possible throughout the construction phase. Established detours with appropriate signing also reduce congestion. Hours of construction limited to off-peak or even nighttime periods limit the effect of traffic delays. A public information system, giving preliminary details on what streets, ramps, etc., will be closed during what time periods, can assist motorists in selecting alternate routes.

9.6.3 Secondary impacts

Mitigation for potential impacts on air quality, noise levels, and visual amenities is discussed in the chapters on those impact categories. These are examples of measures to reduce transportation and access impacts:

- Provision of special access for emergency vehicles
- Construction of frontage roads for access to businesses
- Installation of signals, sidewalks, or other measures to increase motorist and pedestrian safety
- Blocking off of residential streets to prevent through traffic
- Expanded transit service
- Vehicular or pedestrian overcrossings or tunnels
- Incorporation of specific design features (such as bikeways) to enhance bicycle and pedestrian travel
- Coordination with local planning organizations to ensure compatibility with transportation goals and policies

The above examples of mitigation measures for traffic and transportation impacts are certainly not all-inclusive. The analyst should examine both the existing local transportation system and character, and the future changes with the proposed project or action to identify potential impacts. Many projects can have effects at both local and regional levels. Coordination with local, regional, and state agencies will greatly assist in the identification of impacts and the development of feasible mitigation measures.

Chapter

10

Section 4(f)

If the proposed project is a federally funded transportation project, Section 4(f) of the DOT Act of 1966 applies [now Section 303 in Title 49, but still commonly referred to as Section 4(f)]. Because it is not the intent of this text to explain detailed application of all possible impacts for every type of project, the following discussion should be considered basic and introductory. The reader should be aware that Section 4(f) involvement is a very serious limitation to the advancement of a project. If the proposed project is a federally funded transportation project, detailed investigation of Section 4(f) requirements, beyond those presented here, should be undertaken. The best source of detailed information is the Department of Transportation in its regulations and specific guidelines and handbooks.

10.1 Legislation

Section 4(f) basically states that the Secretary of Transportation may approve a project requiring use of publicly owned land of a public park, recreation area, or wildlife or waterfowl refuge, or land of an historic site of national, state, or local significance (as determined by the officials having jurisdiction over the park, recreation area, refuge, or site) only if

1. There is no feasible and prudent alternative to such use.

2. The project includes all possible measures to minimize harm.

10.2 Applicability

Section 4(f) normally applies to public parks; lands included within the National Wildlife Refuge System; recreational, but not scenic, segments of federally designated wild and scenic rivers; national parks;

lands of the National Fish Hatchery System; and any federal or state park or recreation lands acquired under Section 6(f) of the Land and Water Conservation Act.

The applicability of Section 4(f), however, often is not a simple determination. The gray areas of Section 4(f) can become obvious by examining the language of the law; and they usually involve definitions of terms, such as *use, publicly owned, significance,* and *feasible and prudent,* and even in the definition of a park or recreation area.

A few examples can illustrate the types of situations in which it may be difficult to determine Section 4(f) applicability.

Use could mean actual taking of property or use through easements. The concept of *constructive use* also has been applied when the land is not actually touched, but other proximity effects (such as noise, air quality, or visual) intrude into the Section 4(f) property and hinder its use. Temporary construction activity disturbance could be a use, as well as tunneling beneath a potential Section 4(f) property, but only under certain additional circumstances.

A park is determined public under Section 4(f) if the entire public is permitted visitation at any time, but not if the area is limited to only a select group. The purpose and use of an area may also enter into the determination if the land in question is a public park or recreation area. Ambiguous situations arise for school playgrounds, fairgrounds, trails, bodies of water, or bikeways.

Publicly owned lands may include public easements or lease agreements.

Section 4(f) normally applies to any take of land from the parcel in question, unless the official having jurisdiction over the land determines that the entire parcel is not significant. This feature of Section 4(f) is usually the most difficult to explain to the general public or to local public officials. They often will indicate that it's acceptable to take a small piece of their local city park, particularly if the alternative with the small take is preferred by local citizens and officials. It's very difficult to explain that they do not have that option. Section 4(f) will apply unless they determine the entire park to be insignificant. Fortunately, programmatic Section 4(f) statements have now been implemented for use on minor takes of land, as discussed later in this chapter.

Exceptions to the need to declare the entire parcel insignificant can be made, however, for public multiple-use land holdings, such as national parks, state parks, or lands within the jurisdiction of the Bureau of Land Management. If an established management plan does not officially designate the affected portion of the parcel for recreational use (such as may exist with a state forest), then Section 4(f) does not apply. If no management plan exists, then a judgment must be made of the primary function of the land to be affected.

For historic and archaeological resources, Section 4(f) applies to those eligible for inclusion on the National Register of Historic Places. For archaeological resources, Section 4(f) applies only to those resources found significant for preservation in place, and not for data recovery.

For historic bridges and other historic structures, Section 4(f) applies if an adverse effect is determined under the criteria of the National Historic Preservation Act.

Although Section 4(f) applies to wildlife refuges, it may or may not apply to wildlife management areas. For Section 4(f) to apply, the publicly owned wildlife management area (or preserve, reserve, or sanctuary) must perform the same function as a wildlife refuge. If the area functions for the protection of a species, Section 4(f) applies.

The best source of information on uncertain areas of Section 4(f) applicability is the Federal Highway Administration (FHWA) Policy Paper on Section 4(f), which gives specific applicability guidance on a variety of circumstances (1989).

10.3   Avoidance Alternatives

Alternatives to avoid Section 4(f) involvement must be developed, if none already exist. In most cases, the alternative that avoids the Section 4(f) property must be selected. Often the avoidance alternatives are, in fact, very feasible. Alternatives determined not feasible and prudent must meet very stringent requirements of extreme magnitude of impact, or of not meeting the defined purpose and need of the project. As previously discussed in Chap. 4, Alternatives, the initial definition and identification of project purpose and need are extremely important, especially when one is deciding to eliminate an alternative from consideration as not feasible or prudent.

The extreme importance of Section 4(f) makes it essential to identify potential Section 4(f) involvements early in the environmental impact assessment process, preferably at the overview stage. Because alternatives to avoid the involvement must be developed, and at an equal level of detail as other alternatives, it is clear how a late discovery of Section 4(f) involvement can have disastrous effects on project schedules and budgets.

10.4   Format and Processing

Compliance with Section 4(f) is called a *Section 4(f) Evaluation*. It can be a separate document, for example, for minor projects, but it is normally included within the Environmental Assessment (EA) or Environmental Impact Statement. The draft Section 4(f) Evaluation is included in the Draft Environmental Impact Statement or the Environmental Assessment. The final Section 4(f) Evaluation becomes a part of the Final Environmental Impact Statement or the Finding of No Significant Impact (FONSI).

The Section 4(f) Evaluation should include the following:

1. Description of Section 4(F) property
   a. Size, access, and location
   b. Function and available activities
   c. Facilities existing and planned use
   d. Relationship to similarly used lands
   e. Ownership
   f. Unusual characteristics
2. Impacts
   a. Size of area taken versus that of remaining lands
   b. Facilities to be removed or affected
   c. Intrusions of noise, air pollution, visual impacts
3. Avoidance alternatives
4. Mitigation measures
   a. Functional replacement
   b. Construction
   c. Replacement land
5. Coordination
   a. Department of Interior
   b. Department of Agriculture (if appropriate)
   c. Department of Housing and Urban Development (if appropriate)
   d. Responsible officials

If circulated as a separate document, the Section 4(f) Evaluation also should include sections on the project's purpose and need and a project description. If contained and circulated within an EA or Draft EIS, the Section 4(f) Evaluation does not need to repeat the alternatives section of the DEIS, unless there are no proposed alternatives that avoid use of Section 4(f) land or unless such alternatives have previously been eliminated from consideration. For alternatives previously eliminated, the Section 4(f) Evaluation portion of the DEIS must explain the reasons why those alternatives were deemed not to be feasible and prudent.

10.5   Final Section 4(f) Evaluation

A final Section 4(f) Evaluation can be prepared as a separate document or can be included in the FONSI or FEIS. The Final Section 4(f) Evaluation must document the basis for concluding that there are no feasible and prudent alternatives to the use of Section 4(f) lands. To do this, it must be shown that there are unique problems or that the cost, social, economic, and environmental impacts or community disruption would reach extraordinary magnitudes.

The final Section 4(f) Evaluation must discuss the basis for concluding that the proposed project or action includes all possible planning measures to minimize harm to the Section 4(f) property. If the selected alternative uses land from a Section 4(f) property, that alternative must be shown to cause the least harm, after mitigation, to the Section 4(f) resource.

The final Section 4(f) Evaluation must document coordination and comments received during circulation of the draft.

10.6   Programmatic Section 4(f) Evaluations

Because many small projects can involve similar minor takes of Section 4(f) property, programmatic Section 4(f) Evaluations have been prepared to apply nationwide to appropriate minor take projects. The application of programmatic Section 4(f) Evaluations can save time for minor projects not requiring major environmental documents. The use of a programmatic Section 4(f) Evaluation is not a waiver or an exemption from compliance with the law, but is the actual Section 4(f) Evaluation for projects where preset conditions apply. When projects meet all these preset conditions, then the programmatic Section 4(f) Evaluations can be used.

Programmatic evaluations have been developed for minor takes of public parks, recreational lands, and wildlife and waterfowl refuges; minor use of historic sites; and necessitated use of historic bridges. These are some of the criteria of applicability:

- Improvements have been made on same location.
- Section 4(f) property is adjacent to highway.
- Use of the land does not impair use of remaining land.
- Proximity impacts do not impair use of remaining land.
- Agencies and the owner of the land must agree to land conversion or transfer.
- Official having jurisdiction must agree with the assessment of impacts and proposed mitigation.
- Alternatives of do nothing, avoid use of land, and build on new location must have been considered and found not prudent or feasible.
- Project does not remove or alter historic buildings, structures, or objects.

- Project does not disturb archaeological resources that are important to preserve in place.

- Project has no effect or no adverse effect under Section 106 of the National Historic Preservation Act.

- Sponsoring agency must affirm that all possible measures to minimize harm are incorporated into project planning.

11

Energy

In the 1970s, the United States was characterized by long lines at gas stations. Based on vehicle license plate numbers, citizens were not permitted to buy gas on particular days. U.S. residents woke up to an energy crisis and the realization that energy resources can indeed be exhausted. During those times of energy shortages, environmental impact assessment guidelines were revised to reflect the national concern. An Environmental Impact Statement had to contain a separate chapter discussing in detail the potential impact of energy consumption of the proposed project or action. As times change, so does the emphasis placed on particular areas of impact evaluation. The assessment of energy consumption impacts, however, remains a part of an analysis of potential impacts of proposed projects or actions. The appropriate level of detail will differ greatly with the type and magnitude of project or action being evaluated.

11.1 Terminology

Energy is the capacity of a physical system to do work. We refer to energy resources predominantly as the fuels and methods used to produce energy. Fuels may include oil, natural gas, coal, wood, radioactive materials, sunlight, or wind and water in the sense that rapid movement through turbines or windmills can generate electric energy. In this sense, energy is comparable to the term *power.*

The common measurement unit for energy consumption is the British thermal unit, abbreviated Btu. One Btu is the amount of heat required to raise the temperature of one pound of water one degree Fahrenheit.

11.2 Proposed Project or Action Characteristics

The first step in the assessment of potential energy impacts is a thorough description of relevant features of the proposed project or action. The description should include the following:

- All energy-consuming equipment and processes which will be used during construction and operation of the project
- Consideration of the total lifetime of the project or action, if possible
- Energy efficiency of required materials, fuels, or equipment
- The potential of the project to generate trips, including the number of trips, mode of transportation (car, truck, bus, transit, or rail), length of trips, and use of fuel
- Other possible secondary effects of the proposed project or action that may cause energy consumption, such as related facilities or population growth
- Any energy conservation features, equipment, or incentives
- An estimate of the total energy requirements based on fuel consumption or other energy use

The Affected Environment section of the environmental document should identify suppliers of electricity, gas, and other energy fuels and sources, including capacities and related generating facilities.

11.3 Impact Analysis

Most all activities consume energy in some form or another. The goal of the environmental impact analyst is to determine the degree to which a particular proposed project or action consumes or conserves energy or fuels compared with a no-action, or no-build, alternative. The assessment may be no more than a paragraph or two, qualitatively describing the expected impact of the project. A more detailed quantitative analysis may be required for more complex projects, or for projects or actions that directly involve the use or production of energy or energy resources, such as nuclear power plants, hydroelectric dams, coal mining, methane recovery, development of solar-powered vehicles, windmill construction, and oil and gas drilling.

Consumption of energy can have a direct or an indirect impact on the proposed project or action. Energy consumption or conservation effects can be short-term or long-term and, if possible, should be calculated over the lifetime of a project or action. Energy use also is the type of impact included in the section of the environmental document related to irreversible and irretrievable commitments of resources. The discussion may include an assessment of how the project may preempt future energy development or future energy conservation.

Impact assessment between no-action and proposed action conditions, and among various proposed alternatives, can be made by comparing actual units of the particular resource, or by converting all energy resources to British thermal units. For example, the electricity use of various alternatives can be directly compared in units of kilowatthours per day or gigawatthours per year. Natural gas consumption can be compared directly in thousands of therms per year.

For alternatives involving differing energy source usage, conversion to Btu's permits a direct comparison. Conversion factors for kilowatthours, therms, gallons of gasoline or diesel fuel, and gallons of crude oil are available from U.S. Department of Energy publications. For example, 1 kWh of electricity is equal to approximately 3414 Btu, and the heat of combustion of 1 gal of diesel fuel is equal to approximately 147,600 Btu (LACMTA 1993). Selected proposed action alternatives can thus be compared to each other and to the no-action alternative.

A light-rail transit project may use electricity to power the trains and at the stations, increasing the total consumption of that resource. The associated effect of possibly reducing trips in individual vehicles as riders are diverted to the transit system must then be entered into the assessment as an offsetting factor. Energy saved through fewer vehicle trips or better flow of traffic should be calculated by estimating corresponding reductions in consumption of gallons of gasoline or diesel fuel. If an all-bus alternative is being considered, the energy consumption of that alternative must be changed to Btu's from diesel fuel, or from compressed natural gas (CNG) if the system uses CNG buses.

Energy consumption of a no-action, or no-build, alternative may consist of existing conditions or calculations of future without-project conditions. The energy ramifications of not taking action or not proceeding with a proposed project can frequently be overlooked, but should, in fact, be an important consideration in the overall analysis. A proposed action or project may provide improved energy or fuel use efficiency, less reliance on nonrenewable resources, or greater opportunities for energy conservation than a no-action alternative.

To properly calculate energy use, the assumptions and information on the proposed project must be used. For example, to estimate the use of energy for a highway project first use information and/or assumptions on direct energy consumed by vehicle propulsion, including total vehicle miles traveled, mix of trucks and passenger cars, traffic volumes, speeds, delays at signals, distance traveled, and thermal value of the utilized fuel. Next, gallons of gasoline or diesel fuel consumed per mile of travel must be estimated. Finally, energy consumption for construction activities must be included.

A specific example of the calculation of energy consumption of a proposed highway project (U.S. DOT, FHWA, and PADOT 1993) uses the procedure summarized below.

Step 1.   Calculate energy factors for automobiles, medium-weight trucks, and heavy trucks by multiplying the percentage of each vehicle type times a vehicle energy factor based on average assumed speed for each alternative and for the no-build alternative. Add the resultant three numbers and divide by 100 to get a *composite vehicle energy factor.* (It will be a number such as 4.034.)

Step 2.   Multiply the composite vehicle energy factor from step 1 by the average daily traffic by the length of highway by 10^6, to get the *total annual vehicle energy,* in Btu's, for the year of completion and for the year of completion plus 20 years, such as 2001 and 2021, for the no-build alternative and each proposed build alternative. (It will be a number such as 3.46×10^{10} Btu.)

Step 3.   Multiply the construction energy factor (from known projects) by the construction cost of each alternative by 10^4, to get the *total construction energy.* (It will be a number such as 9.46×10^{10} Btu.)

Step 4.   Add the 2001 and 2021 average daily traffic values, and divide by 2 to get an average for the 20-year period. Multiply that average by the composite vehicle energy factor by the number of miles saved, compared to the no-build alternative, by 10^6, to get the *energy savings per year* over the 20-year period for each build alternative. (It will be a number such as 3.39×10^{10} Btu/yr.)

Step 5.   Add the total annual vehicle energy for the year of completion (2001) and the total construction energy, then divide by the energy savings per year to determine the *recovery period* for each build alternative. (It will be a number such as 3.46×10^{10} plus 9.46×10^{10}, divided by 3.39×10^{10}, or 3.8 years.)

This example is given to illustrate the numerous steps in an energy analysis and the variety of required input factors.

The efficiency or wastefulness of project machinery, fuel use, induced transportation, or energy resource requirements should be discussed in the environmental document. Related environmental effects of various energy uses also may be a consideration. Certain types of fuels are known to be more efficient and to produce less air pollution, or disturbance to natural resources, than other types of fuels. Whether the project uses renewable or nonrenewable energy, resources will be of interest to resource agencies and the public.

The ability of existing energy suppliers, such as local electric companies, to meet the needs of the proposed project or action can be determined through coordination with local utilities and comparison with existing capacities.

Results of an analysis of energy effects should comparatively evaluate all alternatives in terms of overall energy consumption, and in terms of reducing wasteful, inefficient, and unnecessary consumption of energy during project construction, operation, or maintenance. Criteria for assessing significance, or degree, of impact may include the following:

- Overall energy consumption

- Wastefulness or efficiency of energy use

- Consumption during construction or operation beyond capacity of energy suppliers to meet demands

- Need for construction of additional facilities for energy generation or distribution to meet increased demand.

11.4   Secondary and Cumulative Effects

The fact that changes in energy consumption often have ripple effects, or secondary ramifications, can lead to an estimated impact analysis at best. Many, many diversified energy resources and energy consumption uses can be indirectly related to a proposed change in use of one resource. Another consideration, especially for land-use and management plan assessments, is the potential for a proposed action to place limitations on energy and mineral exploration and development. The impact of such a restriction often cannot be quantitatively determined.

11.5   Mitigation

The study should explore the feasibility of measures to reduce energy consumption, increase energy efficiency and conservation, reduce any associated environmental effects of energy use, and reduce dependency on nonrenewable resources. Mitigation measures may include these:

- Incorporation of energy-efficient equipment and practices into construction and operational procedures

- Reduction of wasteful, inefficient, and unnecessary consumption of energy

- Use of location, orientation, and design features to minimize energy consumption, including transportation energy

- Timing of energy use to not coincide with peak energy demand periods
- Use of alternate fuels or energy systems
- Energy conservation through recycling
- Recovery and use of by-products or wastes of the process to produce energy, such as methane recovery from landfills

As with other areas of potential environmental effects, the assessment of possible energy impacts and mitigation measures should be based on an inquiring thought process that can include possibilities related to project design, operation, and conservation options. The awareness of energy as a resource, and the examples known from other projects or experiences, will lead to an appropriate consideration of possible energy effects of the proposed project or action.

12

Historic and Archaeological Resources

In compliance with Section 106 of the National Historic Preservation Act (NHPA) of 1966, as amended, and other related legislation and regulations, the identification and impact evaluation related to historic and archaeological resources must follow a stringent, systematic, timely procedure. This chapter examines what is commonly referred to as the *Section 106 process*. The goal here is to present a basic understanding of the requirements and the process. Several excellent, more detailed, publications are available from the Advisory Council on Historic Preservation (ACHP), an independent reviewing agency established by the National Historic Preservation Act.

The environmental impact analyst should not become overwhelmed by the numerous individual laws and regulations related to historic and archaeological resource protection. It is not necessary to be familiar with every statute. By following the direction given within the ACHP regulations, "Protection of Historic Properties" at 36 CFR Part 800, and other ACHP user-friendly publications, the analyst either will already be in compliance with pertinent other laws and regulations or will be referred to them as appropriate. This chapter refers only to those laws or Codes of Federal Regulation (CFR) which are generally in more common usage and should be familiar to even the nonhistorian environmental assessment analyst.

Although the laws and regulations are the same, the environmental impact analyst should be aware that requirements for Section 106 vary immensely from state to state. What may be an appropriate level of effort to meet requirements for surveys in one state, or even agency, may not be the "appropriate" level of effort in a different state or for a different agency. Required documentation in the form of specific reports and forms for each step of the process also may vary significantly by state. The analyst should be aware that each state or agency sincerely believes that its way is not only the best approach, but the *only* appropriate approach. Many times there are legitimate reasons for these geographic or agency peculiarities.

This chapter points out some of the geographic and agency variations that the author has experienced. The analyst, however, should become thoroughly familiar with the acceptable methodologies for his or her particular state, agency, or project being assessed.

As with other chapters in this text, this description of the requirements of Section 106 procedures is not intended to replace the inclusion of a qualified local historian and archaeologist on the study team. Whereas evaluation of potential project effects may be conducted by more of a generalist, the identification and description of resources must be completed by qualified personnel with the appropriate educational and experience backgrounds.

12.1 Comparison of Section 106 and Section 4(f)

Section 106 of the National Historic Preservation Act applies to federal, federally assisted, or federally licensed and permitted projects and actions. Its protection is afforded to properties listed on, or determined eligible for, the National Register of Historic Places (NR). The National Register of Historic Places is contained within 36 CFR Part 60 and does not necessarily contain only properties or sites of national significance, but also of state or local significance. The Section 106 process has two basic requirements: (1) The sponsoring agency must take into account the effect of the proposed undertaking on any district, site, building, structure, or object that is included in, or eligible for, the National Register of Historic Places. (2) The Advisory Council on Historic Preservation must be afforded a reasonable opportunity to comment.

Section 110 of the National Historic Preservation Act provides direction for historic preservation management within federal agencies for resources owned or controlled by the agencies. Section 110 also provides protection to National Historic Landmarks (so designated under the authority of the Historic Sites Act of 1935), as described in the ACHP regulations, "Protection of Historic Properties," 36 CFR Part 800. National Historic Landmarks are historic properties of outstanding national significance that have been specially designated by the Secretary of Interior. Under Section 110 of the National Historic Preservation Act, National Historic Landmarks require special treatment: Agencies must "to the maximum extent possible, undertake such planning and actions as may be necessary to minimize harm to any National Historic Landmark that may be directly and adversely affected by an undertaking." This special consideration applies in addition to requirements of Section 106.

Section 4(f) of the DOT act of 1966 also applies to an adverse effect on a National Register listed or eligible historic or archaeological property, or a taking of that property, for federally funded transportation projects (refer to Chap. 10). An exception is that Section 4(f) does not apply to archaeological resources significant for the recoverable data contained, but only to archaeological resources significant for preservation in place.

Of the two legislative requirements, Section 4(f) is the more restrictive. Section 4(f) explicitly states that the transportation project cannot be approved unless there is no feasible or prudent alternative. *Feasible* and *prudent* do not mean just convenient, but that the avoidance alternative would have to cause impacts of an extreme magnitude. For example, the taking of residential homes to avoid Section 4(f) involvement is often considered completely within the realm of feasible and prudent. Section 4(f) absolutely prohibits selection of an alternative involving Section 4(f) property, if another alternative is deemed to be feasible and prudent.

The Section 106 process does not require preservation of every historic property in every case. The Advisory Council on Historic Preservation does not have veto power over an agency's action. The sponsoring agency has the ultimate right to make the final decision, even if it disagrees with opinions of the State Historic Preservation Officer or the Advisory Council on Historic Preservation. Results of the Section 106 process can range from full preservation to unmitigated destruction of a property. The process does ensure that an agency weighs preservation into the balance with the projected benefit of the completed undertaking, costs, and other factors.

12.2 Timing of Section 106 Activities

The Section 106 process includes numerous steps, as outlined in Fig. 12.1. Simplified, the three basic steps are as follows: (1) Identify resources; (2) assess impacts; and (3) if an adverse effect is identified, consult about ways to avoid, reduce, or mitigate the harm. These three steps are about the same as those undertaken for any area of potential environmental impact. Although the three basic steps are identification, assessment, and mitigation, there are numerous intermediate requirements within each basic step for compliance with the Section 106 process.

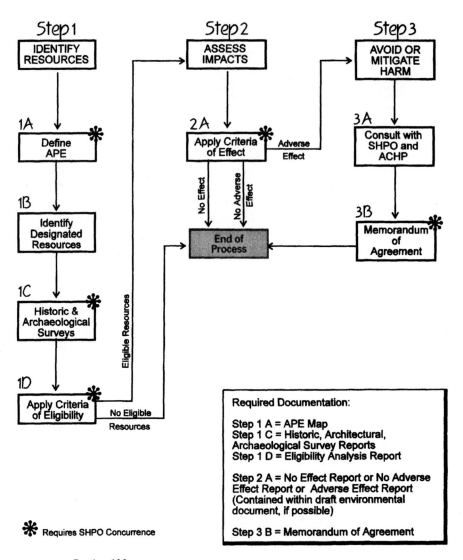

Figure 12.1 Section 106 process.

Although it is a separate legal requirement, Section 106 processing should be coordinated as much as possible with the environmental review process of NEPA. Because a public review period is required for the Section 106 process, it is extremely desirable to include the possible project effects on historic and archaeological resources within the NEPA required Environmental Assessment or Draft Environmental Impact Statement. Because there are so many steps to the process prior to determination of impacts, and many of these initial steps require coordination with, and concurrence by, other agencies, it is imperative that historic and archaeological studies begin immediately in the study process.

For example, concurrence of the State Historic Preservation Officer is required on the area of potential effect, the eligibility of potential resources for the National Register of Historic Places, and the results of application of the criteria of effect. All this needs to be accomplished if the Environmental Assessment or Draft Environmental Impact Statement is to include a description of impacts. The surveys needed just to gather the information required to initiate consultation for these concurrences can be extensive. If a review time of, say, 4 to 6 weeks is factored into the process for the State Historic Preservation Officer to respond, it is easy to understand how the Section 106 process can severely affect the schedule and progress of the draft environmental document.

Section 106 process documents can, however, be circulated for review separately from NEPA environmental documents, if it is not possible to include them within the Environmental Assessment or Draft Environmental Impact Statement. The final NEPA environmental documents, however, should document the outcome of the Section 106 process. Because it is a separate legal requirement, Section 106 compliance is applicable to undertakings that do not require an Environmental Assessment or Environmental Impact Statement under NEPA. Section 106 prohibits expenditure of public funds on a project until the process is successfully completed.

Consideration of historic and archaeological resources must occur in the early stages of project planning so that preservation concerns can receive thorough consideration as a project or action is planned. Early preservation review also permits modification to a project while such modifications are relatively easy to accomplish and reduces the potential for conflict and delay.

12.3 Area of Potential Effect

While in many states this early requirement of the Section 106 process is ignored, at least in the formal sense, in other states it is considered a major task. The task is to define the boundary of the possible impacts of the proposed project or action that may cause changes in the character or use of historic properties. This boundary then sets the geographic limits of subsequent historic and archaeological resource surveys.

The *area of potential effect* (APE) always includes all lands to be disturbed or cleared by the proposed project or otherwise affected by a proposed action. The APE also normally includes the area immediately adjacent to the area to be physically disturbed. The definition of the APE must consider possible direct, indirect, secondary, and cumulative impacts for all proposed alternatives. It therefore does not necessarily need to be contiguous. For example, if alternatives include several different geographic sites, an APE can be established for each alternative site.

It is easy to recognize one of the inherent problems in integrating the Section 106 process into the NEPA process. Because of the extensive, and often time-consuming review requirements of Section 106, it is imperative that the process begin very early in the environmental impact studies. The definition of boundaries of the APE, however, must incorporate possible direct, indirect, and secondary effects of other disciplinary studies, such as traffic, population changes, air quality, noise, or visual impacts. If the Section 106 process begins immediately upon study initiation, these other areas of impact analysis normally have not even begun, *let alone* have conclusive results available.

The APE should be defined as early in the environmental analysis process as possible. Coordination with and concurrence of the State Historic Preservation Officer (SHPO) and the state archaeologist on the boundaries of the area of potential effect are required. Local historians, historical societies, and universities also should be consulted.

12.4   Description of Previously Identified Resources

The identification of resources within the area of potential effect will begin with examination of available material to determine the presence of existing designated resources. Section 106 applies to properties on, or eligible for inclusion on, the National Register of Historic Places. The National Register of Historic Places, published in the *Federal Register,* should be reviewed for the properties or sites within the area of potential effect. Information will include those properties listed on the National Register of Historic Places and a separate list of those properties determined eligible for the National Register of Historic Places but not actually on it.

The State Historic Preservation Officer will have valuable information on properties listed on the National Register of Historic Places and on previous surveys conducted in the state or project area. Other available sources of information include local governments, local historians and historical societies, and books and surveys on the history and archaeology of the area. Local colleges and universities are always a good source of information. Any previously prepared Environmental Impact Statements for other projects or actions within the study area should always be reviewed for information related to historic and archaeological resources, as well as other potential areas of impact.

The public participation and scoping process established for the proposed project or action should include opportunities for the identification of sources of information and for individuals to express their concerns regarding historic properties early in the agency's planning process, so concerns can be considered in a timely manner.

Section 106 specifically requires consideration of special social and cultural values, particularly religious concerns, of Native American groups (American Indian Religious Freedom Act). The regulations require that representatives of Native Americans be brought into the consultation process when historic properties of importance to them may be affected. If an APE includes lands owned by, or of interest to, Native Americans, local tribe leaders should be consulted for information.

A list of known designated resources should be prepared, noting the status of the resources, such as on or determined to be eligible for the National Register of Historic Places, identified as having value in local surveys, or designated as a National Historic Landmark.

After previously identified resources have been located and described, the area of potential effect must be surveyed for any unknown or unidentified resources. This requirement is rather unique in some ways to the Section 106 process. Section 106 protects properties on the NR and identified as eligible for the NR, but also properties not previously identified as eligible. It is the sponsoring agency's responsibility to determine if any previously unidentified eligible properties may exist within the area of potential effect.

This requirement is noted as somewhat unique because when compared with protective legislation and regulations in other areas of potential impacts, it is normally not the responsibility of the sponsoring agency to identify previously unknown resources for protection under the law. Examples include the Endangered Species Act, wild and scenic rivers, or public parkland. In these cases, the agency is responsible for determining whether any designated resources exist within the project area, but not for conducting the required research to determine if, for example, a species should be listed as threatened or endangered, if a river should be included as wild and scenic, or if a park should be established. There are exceptions, of course, when information and public or resource agency sentiment are extremely indicative of pressing for such incorporation of important, previously nondesignated species, rivers, or open space under the protection of applicable laws.

As opposed to endangered species and other examples listed above, the Section 106 process does place responsibility on the sponsoring agency to do the required research to determine if any previously unidentified resources exist within the APE and whether those resources meet the criteria of eligibility for the National Register of Historic Places.

12.5 Historic and Archaeological Resource Surveys

The objective of historic and archaeological resource surveys is to identify all properties that may be eligible for the National Register of Historic Places. The survey should be conducted by qualified historians and archaeologists. Local knowledge of the area is useful to save time and to ensure that all resources are identified.

12.5.1 Eligibility for the National Register

National Register eligibility is determined through application of specific criteria. The National Register of Historic Places is the official list of properties significant in U.S. history, architecture, archaeology, engineering, and culture. The criteria to evaluate significance used by the historic and archaeological teams to determine eligibility are contained within 36 CFR Part 60. As with other legislation and regulations, there are numerous helpful publications regarding application of the criteria. Basically, the criteria are as follows.

The quality of significance in U.S. history, architecture, archaeology, engineering, and culture is present in districts, sites, buildings, structures, and objects that possess integrity of location design, setting, materials, workmanship, feeling, and association and

- That are associated with events that have made a significant contribution to the broad patterns of U.S. history
- That are associated with the lives of persons significant in our past
- That embody the distinctive characteristics of a type, period, or method of construction, or that represent the work of a master, or that possess high artistic values, or that represent a significant and distinguishable entity whose component may lack individual distinction
- That have yielded, or may be likely to yield, information important in prehistory or history

Ordinarily cemeteries, birthplaces, or graves of historical figures; properties owned by religious institutions or used for religious purposes; structures that have been moved from their original locations; reconstructed historic buildings; properties primarily commemorative in nature; and properties that have achieved significance within the past 50 years are not considered eligible for the National Register. Such properties can qualify, however, if they meet certain special criteria.

Specific guidelines are available to the historian or archaeologist in applying the criteria and in the definition of the terminology. The most helpful perhaps is published by the Department of the Interior, "How to Apply the National Register Criteria for Evaluation." These guidelines include the following.

Categories—how to define the categories of historic properties as districts, sites, buildings, structures, or objects.

Historical context—eligibility of a property depends on whether it represents a significant theme or pattern. Decisions on significance, thus, can reliably be made only within the context of an area's history, and guidance is available on how to define historical context, a first step in designing a field survey.

Level of significance—guidelines for classification of importance of a property as having local, state, or national significance.

Type of significance—application of the four specific criteria (A, B, C, or D) and determination of which applies within the context of identified relevant historical themes or patterns. Specific guidelines and examples of eligible and noneligible properties are given for each criterion.

Integrity—seven criteria that are listed and explained: location, design, setting, materials, workmanship, feeling, and association. Guidelines are given with examples of eligible and noneligible properties.

Criteria for properties normally not considered eligible—separate guidelines and criteria for religious properties, moved properties, birthplaces or graves, cemeteries, reconstructed properties, commemorative properties, and properties less than 50 years old.

12.5.2 Historic resource surveys

The survey of aboveground resources often begins by the process of elimination of recent structures, or properties that normally are not considered eligible for the National Register of Historic Places. There is usually an agreement with the State Historic Preservation Officer of a particular state on the cutoff date for structures possibly eligible for the National Register of Historic Places. The acceptable age of structures is normally at least 50 years, but can be less if special circumstances are involved which would give the resource exceptional historic value.

As noted earlier, each state will likely have its own required methodologies, forms, and documents for the conduct of surveys. Many states use the actual National Register of Historic Places nomination forms, even though the properties are not actually being nominated for the listing.

The historian or historical architect surveys all aboveground structures within the area of potential effect and prepares the appropriate forms for each property. The accompanying report normally separates properties thought not to be eligible, properties thought to be eligible, and properties for which eligibility is uncertain. It is the sponsoring agency's responsibility to determine National Register eligibility. The Determination of Eligibility report, or in some cases numerous reports of different titles, then facilitates coordination with the State Historic Preservation Officer for review and concurrence on the results of the survey.

12.5.3   Archaeological surveys

The identification of prehistoric and historic archaeological resources begins with research of available information on the area within the APE and the context of the region. The state archaeologist should be consulted for appropriate known sites and previous surveys. Topographical maps of the area can be reviewed for possible indication of resources.

Perhaps more than aboveground resources, appropriate methodologies and formats of documentation for archaeological studies vary immensely from state to state. Because of this great flexibility in implementing field surveys, it is important that selection of techniques and level of effort be responsive to the specific goals of the results. Surveys can range from simple "windshield" or walk-over surveys with very limited subsurface examination, often referred to as *reconnaissance surveys,* to extensive, subsurface investigation to determine precise information on the property, including significance, integrity, and boundaries sufficient to permit an evaluation under criteria of eligibility for the National Register of Historic Places.

The philosophy of archaeological resource preservation often differs from that of aboveground historic resources. Extremely oversimplified, this philosophy involves the desirability of leaving archaeological resources intact for future generations, with better methods, to explore. In other words, do not disturb what might be there just to find out in detail and recover what might be there. This philosophy is rooted in past experiences when extremely sensitive and valuable archaeological resources were extensively excavated with technology of the times; sites were thus destroyed for future technologies that would have yielded significantly more data.

In most cases, the minimum information necessary to evaluate properties against National Register of Historic Places criteria of eligibility should direct the extent of information-gathering activities. Resources should be evaluated within particular historical context and should be classified in terms of type and integrity. The location and physical extent of the property should be determined. It also is very important to determine if the archaeological resource is valuable for the information it may contain, as opposed to being significant for preservation in place.

The level of survey for archaeological resources should be determined in consultation with the State Historic Preservation Officer and the state archaeologist. Most states have an established methodology for survey types and techniques under various proposed project or action circumstances. As with aboveground resources, archaeological resources survey information is available in several helpful publications from the Department of the Interior.

It is most important to remember that surveys should be appropriate in scale to the purpose of the evaluation. If the goal is to recover any discovered archaeological resources and artifacts from the site, then data recovery programs for the entire site may be conducted prior to any construction activities. On the other hand, an assumption of National Register eligibility may be agreed upon with minimum disturbance of a site, and the process continued directly to developing mitigation (Memorandum of Agreement) techniques to ensure resources are identified and recovered if found during actual construction activities.

For example, a proposed new bridge will require excavation on river banks for construction of bridge abutments. Research and coordination with the State Historic Preservation Officer indicate a high probability of presence of significant archaeological resources in the location of the proposed bridge abutment. The State Historic Preservation Officer refuses, however, to agree to any unnecessary subsurface disturbance of the site to determine in detail what may be there and what the exact physical boundaries may be. The property is assumed eligible for the National Register of Historic Places, and the Section 106 process proceeds to development of a mitigation plan. During construction within this particular area, a qualified archaeologist will be present to observe the activities. Should any artifacts be found, the artifacts will be appropriately recovered, but the extent of disturbance will be kept to an absolute minimum.

12.5.4 Classes of historic properties

If a proposed undertaking will have possible effects that are extremely difficult to define or that extend over a large geographic area, agencies may identify and consider *classes* of historic properties. Examples may include large reservoir projects, or housing projects where it may not be feasible to identify all individual properties possibly subject to effect, particularly secondary effect, prior to project approval. In these cases, it is acceptable to predict that the undertaking will affect certain kinds of historic buildings or archaeological sites. Knowing such effects will occur will assist in developing programs to protect the significant characteristics of such properties. Thus, although it is not feasible to identify specific historic properties, the agency has met its responsibilities under Section 106.

12.5.5   Conclusion of identification and eligibility evaluation phase

Results of the identification of historic and archaeological resources within the area of potential effect and the evaluation of those resources for eligibility for the National Register of Historic Places are forwarded to the State Historic Preservation Officer for concurrence. The format for this documentation may be one report or several reports of very differing titles, depending on the state and/or agency involved in the project. The documentation should clearly indicate the sponsoring agency's opinion on whether each historic and archaeological resource or site is considered eligible for the National Register of Historic Places, or considered not eligible, and should make note of determinations which are not absolute and could be decided either way.

If the State Historic Preservation Officer agrees with the agency, the process continues. If the SHPO fails to respond, concurrence is assumed. If the SHPO disagrees about eligibility, the agency must obtain a formal determination from the Keeper of the National Register in Washington, D.C.

If the SHPO concurs that no National Register–eligible historic and archaeological resources are located within the APE, interested parties must be notified and documentation must be made available to the public (within the Environmental Assessment or Draft Environmental Impact Statement if integrated into the NEPA process). Following such notification, the Section 106 process is completed.

12.5.6   Public disagreement

Now is a good time to emphasize the public's right to disagree with findings, even if the sponsoring agency and the State Historic Preservation Officer agree. At four basic points within the Section 106 process, any person, regardless of her or his formal involvement in the process, can request the Advisory Council's review of an agency's findings:

1. Identification of historic and archaeological properties
2. Evaluation of significance of properties
3. Finding that no historic properties are present
4. Finding no effect on historic properties

The council will complete its review within 30 days of the request. The request for review does not require the agency to suspend action on an undertaking. When an inquiry concerns an agency's judgment about National Register eligibility, the matter is referred to the Secretary of the Interior (who refers it to the Keeper of the National Register). There have been cases where the Keeper has overruled the opinions of the agency and State Historic Preservation Officer and has determined sites eligible for the National Register on behalf of a request from the public.

12.6 Impact Assessment

Criteria to determine potential impacts of a proposed project or action on historic and archaeological resources are contained in the Advisory Council's regulations, 36 CFR Part 800. There are two sets of criteria: criteria to determine whether there is any effect at all on the property, either beneficial or adverse; and criteria to determine whether there is an adverse effect on the property.

An effect occurs if the proposed project or action will in any way alter the characteristics of the property that qualify it for inclusion in the National Register of Historic Places.

An adverse effect occurs if the proposed project or action may diminish the integrity of the property's location, design, setting, materials, workmanship, feeling, or association.

Adverse effects include, but are not limited to:

1. Physical destruction, damage, or alteration of all or part of the property;
2. Isolation of the property from or alteration of the character of the property's setting when that character contributes to the property's qualification for the National Register;
3. Introduction of visual, audible, or atmospheric elements that are out of character with the property or alter its setting;
4. Neglect of a property resulting in its deterioration or destruction; and
5. Transfer, lease or sale of the property.

There are exceptions to the above criteria of adverse effect contained within 36 CFR Part 800; perhaps the most notable arises when the historic or archaeological property is of value only for its potential contribution to research and such research is conducted in accordance with applicable professional standards and guidelines.

One of the key factors in assessing whether a proposed project or action will have an effect or an adverse effect is the focus on the characteristics of the property that qualify it for the National Register. Eligibility of a particular property was based on its contribution to a theme or historical context. It may be eligible on the basis of architecture, engineering, ethnic heritage, art, agriculture, invention, or several other categories of significant themes. To properly assess potential project impacts, it is essential to review the data on each property for the identified characteristics that qualified it for the National Register. Remember that to have an effect, the proposed project or action must cause changes in that particular characteristic.

Sometimes a proposed project or action changes the area surrounding the actual building or structure, but does not alter the structure. Boundaries of sites as described on the National Register often coincide with property boundaries of the parcel of land, usually as a matter of convenience. Again, the criterion regarding changes to setting applies in cases of isolation of character when that character contributes to the property's qualifications to be on the National Register. If the project being evaluated will produce physical changes near an eligible site, it is first necessary to examine how much of the surrounding setting actually contributes to the specific characteristics that qualify the property for the National Register. If the proposed project or action will produce changes within an area that actually do not contribute to the value of the resource, then a no-effect or no-adverse-effect determination is appropriate.

Results of the investigation of project effect will be one of three conclusions: no effect, no adverse effect, or adverse effect. Results and supporting documentation are forwarded to the State Historic Preservation Officer for concurrence with the conclusions. For an agreement on a no-effect or no-adverse-effect determination, the Section 106 process is completed after notification to interested parties of its availability to the public (hopefully integrated into the draft NEPA environmental document). A no-adverse-effect finding must also be forwarded to the Advisory Council on Historic Preservation for a 30-day review. If the council disagrees with the no-adverse-effect finding, the effect is considered adverse. A public review period is required, and this review is most efficiently accomplished within the Environmental Assessment or Draft Environmental Impact Statement. The environmental document must contain documentation of agreement of the SHPO on the eligibility of resources and on findings of no effect or no adverse effect.

An adverse-effect finding requires continuation of the Section 106 process to the consultation, or mitigation, step.

12.7   Consultation and Mitigation

The consultation process for a determination of adverse effect is designed to reach agreement on how to avoid, mitigate, or accept the adverse effect. Consultation includes the agency; the State Historic Preservation Officer; the head of local government; owners of affected sites; a representative of an Indian tribe, if the undertaking will affect Indian lands; and other interested persons when deemed appropriate by the agency and State Historic Preservation Officer.

The purpose of the consultation process is to accommodate the needs of the proposed project and the integrity of the historic property in a way that best serves the public interest. The first investigation is usually of alternatives that would avoid or reduce the impact while still accomplishing the agency's goals. Other mitigation measures may include the following:

- Limiting the magnitude of the project
- Modifying the design
- Rehabilitation of an historic property
- Preservation and maintenance operations for historic properties
- Documentation of buildings or structures that must be destroyed or substantially altered
- Relocation of historic properties
- Salvage of archeological or architectural information and materials

In some cases, no alternatives or mitigation is feasible, and the proposed project's benefits in relation to the significance of the historic resource justify destruction of the resource as an acceptable loss. In most cases, however, measures to mitigate impacts are agreed upon; some examples follow.

- Specific design measures will be agreed upon to minimize impacts.
- If it is agreed to destroy an historic building or structure, the usual procedure is to fully document the resource through stringent, acceptable documentation requirements.
- Many times historic buildings can be moved to locations where preservation can be maintained for future generations.
- In cases involving archaeological resources, data can be recovered prior to construction activities, or an archaeologist will observe construction activities and recover data if any artifacts are discovered.
- For the discovery of burials, the area is normally excavated and the remains are moved. If burials are of Native American heritage, an

official of the tribe will most likely be present at the reburial for appropriate ceremony in keeping with tribal traditions.

If a site is to be destroyed because keeping it would result in the adverse-effect criterion of neglect, the site or property should be offered to the local historical society or government if it is willing to assume the cost of maintaining it. This situation often occurs with replacement of historic bridges when the new bridge is built adjacent to the old structure. The Department of Transportation cannot continue ownership and maintenance of the old bridge, but can offer it to the local community.

Results of the consultation process are contained within a Memorandum of Agreement (MOA). The MOA is a legally binding document signed by the agency, State Historic Preservation Officer, and the Advisory Council on Historic Preservation. The signing of an MOA completes the Section 106 process. The public must be afforded an opportunity to review and express its views on an executed MOA. If integrated into the NEPA process, the MOA is included in the final environmental document, either a Finding of No Significant Impact or a Final Environmental Impact Statement.

12.8   Programmatic Section 106 Agreements

In the early years of Section 106 review, the three steps of identification, assessment of effect, and mitigation were almost always applied on a project-by-project basis, and effects were considered for a particular location or site. Under current regulations, agencies may obtain Advisory Council comment on a programmatic basis. As with programmatic Section 4(f) agreements, the programmatic Section 106 agreement is not a waiver of responsibility under the law, but is used in cases where numerous projects are similar (such as replacing historic bridges), an activity covers a large geographic area, or effects cannot be fully determined prior to project approval. The particular proposed project or action must be shown to meet the specific conditions of the programmatic agreement or programmatic memorandum of agreement.

12.9   Summary

As with most regulations, there are exceptions to the process described in this chapter, especially if disagreement occurs at various steps. There are also special considerations not mentioned here, such as when an historic property is designated and discovered late in the project planning and design process. There also is a multitude of ad-

1. Potential historic and archaeological resources identified.	Apply criteria of eligibility, go to 3.
2. No potential resources identified.	Notify SHPO and public, process completed.
3. Property eligible.	Apply criteria of effect, go to 8.
4. Property not eligible.	Coordinate with SHPO, notify public, process completed.
5. Disagreement on eligibility.	Request opinion of keeper of NR, go to 6.
6. Keeper determines property not eligible.	Process completed.
7. Keeper determines property is eligible.	Apply criteria of effect, go to 8.
8. No effect.	Notify SHPO, process completed.
9. Effect determined.	Apply criteria of adverse effect, go to 10.
10. No adverse effect.	Request concurrence of SHPO, go to 12.
11. Adverse effect.	Begin consultation and mitigation process, go to 18.
12. SHPO concurs.	Process completed.
13. SHPO disagrees.	Request council comments, go to 14.
14. Council agrees with no adverse effect.	Process completed.
15. Council recommends changes.	Go to 16.
16. Agency makes changes.	Process completed.
17. Agency does not make changes.	Adverse effect, go to 18.
18. Consultation results in Memorandum of Agreement.	Consult with ACHP, go to 20.
19. Consultation fails, no Memorandum of Agreement.	Consult with ACHP, go to 24.
20. Council signs MOA.	Process completed.
21. Council proposes changes.	Go to 22.
22. Agency makes changes.	Process completed.
23. Agency does not make changes.	Go to 24.
24. ACHP issues written comments, agency considers comments and notifies ACHP of decision.	Process completed.

SHPO: State Historic Preservation Officer
ACHP: Advisory Council on Historic Preservation
NR: National Register of Historic Places

Figure 12.2 Simplified key to Section 106 process.

ditional guidance and regulatory material on the specifics of applying the criteria of eligibility, criteria of effect, and documentation and mitigation techniques. This chapter, however, has presented an overview of the basic steps to a degree necessary for a general knowledge of the Section 106 process and the importance of appropriate timing and coordination during the process. The summary key to activities in Fig. 12.2 may be helpful in understanding and summarizing important steps in the process.

13

Visual Resources

The evaluation of potential visual and aesthetic impacts of a proposed project or action can vary substantially in the level of effort, analysis methodologies, and resultant report documentation. It is important to identify the appropriate scope of study early in the environmental impact assessment process. The selected level of effort should be directly driven by the anticipated magnitude of impact and the importance of obtained results in the decision-making process. Basic steps in visual resources analysis are illustrated in Fig. 13.1.

Many federal agencies have adopted specific guidelines for the evaluation of visual impacts related to the jurisdiction and responsibility of the particular agency. Some state agencies also have guidelines and methodologies to assist the environmental analyst in focusing the visual impact assessment on the needs and possible relevant issues of particular types of projects.

As with many areas of socioeconomic impact analysis, the public perception of visual impacts can be very individualized and subjective. What is beautiful to one person may be unsightly to another. The environmental analyst must maintain objectivity as much as possible in the analysis, and must recognize that there may be a difference of opinion based on individual judgment.

13.1 Inventory of Existing Resources

The description of the existing visual environment will depend on the type of project or action proposed. A site-specific project will require the description of the appearance of the site and surrounding area. Often this description may be divided into foreground, middle-ground, and background descriptions, depending on the characteristics of the

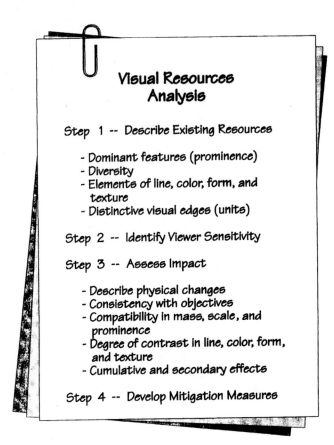

Figure 13.1 Visual resources analysis.

site. All dominant elements of the visual environment should be described, including artificial and natural components.

A good beginning for the analyst is to just tell the reader what the area looks like. This may sound oversimplified, but the analyst must always remember that he or she is writing for the general public. Many of the readers of the environmental document will not have seen the site and, for starters, just need to know what it looks like. The description may begin with a general land-use definition, such as urban, rural, industrial, retail, residential, agricultural, or natural. Natural areas may continue to be described based on land cover, such as forest, woodlots, meadows, streams, mountains, desert, foothills, escarpments, or cliffs. Topography should be described, such as flat, rolling, steep, or mountainous.

Another factor in the description of the existing visual environment is the nature of the viewers. Viewer sensitivity usually relates to the activity taking place and is a measure of the degree of viewer interest in the scenic qualities of the landscape. A hiker in a National Forest will have a very different sensitivity from an urban commuter on the way to work on a congested freeway. The assessment should include an identification of the probable viewers and, in that relevance, the major dominant elements and any significant view sheds or vistas.

If the proposed project or action involves a very large area, such as a proposed community land-use plan, national park, or forest management plan, or a resource recovery proposal over an extensive area, the description of visual resources will, by necessity, be more general. Such an overview should focus on dominant, or particularly significant, features, vistas, or view sheds. Notable visual resources from a regional perspective may already have been identified at a local, regional, or state level. Often these previously identified areas or sites will be noted within land-use, comprehensive, or management plans. Objectives for maintaining resources of significant visual quality also may be included in planning documents.

In an urban setting, the description of the visual environment will cover dominant features and perhaps place more emphasis on elements of line, color, form, and texture. Continuity and contrast can be used to describe visual diversity and interest. Compatibility in architecture, scale, materials, line, and form of urban structures may be important in defining contiguous visual units. Distinctive visual edges between relatively similar areas should be identified.

Many times, the best way to describe the existing visual environment in an environmental document is with the support of drawings or photographs. The analyst should not try to be too subjective in defining visual quality, but rather should describe the components of the existing environment as objectively as possible and let the reader come to conclusions regarding the quality or value of the described environment.

13.2 Examples of Some Federal Agency Methodologies

The Bureau of Land Management (BLM) has a Visual Resources Management Program for the planning and design of the visual aspects of multiple-use land management (U.S. Department of Interior, BLM 1984, 1986a, 1986b). Under the program, Visual Resource Management (VRM) classes of I to IV are assigned to all public lands to describe the acceptable level of change in the landscape and the associated management goals and objectives. The resource management plan would include mapping of areas with each class designation and the total acres within each class. VRM class designations and related objectives are summarized below.

VRM class I has the highest visual resource sensitivity value, with a landscape that appears unaltered by humans. The goal is to preserve the existing landscape with only natural, ecological changes.

VRM class II objectives include maintaining the existing environment with any changes blending into the natural surroundings by repeating the basic elements of form, line, color, and texture found in the dominant elements of the landscape. Changes should not attract the attention of the casual observer.

VRM class III area has a low visual sensitivity resource value. The level of change in the characteristic landscape may be moderate. Management activities may attract attention, but should not dominate the view of the casual observer. Changes should repeat the basic elements found in the predominant natural features of the characteristic landscape.

VRM class IV visual quality objectives provide activities requiring major changes of the existing landscape. These activities may dominate the view and be the major focus of viewer attention (U.S. Department of Interior, BLM 1994).

The above example of the Bureau of Land Management system furnishes meaningful guidance for the analyst. The first lesson is to make full use of work already completed by others. Appropriate research or planning documents, previous Environmental Impact Statements or other studies, and coordination with local or regional officials may result in discovery that visual resources have previously been defined, at least to some degree. Not only does this save the analyst time, but also it is normally a much better assessment than a short-term, project-oriented team could produce.

Second, the BLM system introduces meaningful descriptive terminology of the visual quality inventory and impact process. The objectives refer to lack of change as positive in areas of high visual quality; compatibility with, or a repeating of, the basic elements of form, line, color, and texture found in the dominant features of the landscape; and the ability to be noticed, or prominence of the change to the casual observer.

The USDA Forest Service has a similar system of visual resource management classification, with objectives related to acceptable alteration of the characteristic landscape (USDA Forest Service 1992d). Five levels of visual quality objectives are preservation, retention, partial retention, modification, and maximum modification. The visual resource management objectives in the BLM and Forest Service programs are directly related to an accurate description of the visual quality of the existing landscape.

Use of the Federal Highway Administration's methodology for visual impact assessment, if deemed necessary and desirable based on potential project impact and anticipated degree of project controversy, requires significantly greater detail in describing the existing environment (U.S. Department of Transportation, FHWA 1986). The methodology takes several steps to define the existing visual environment.

The FHWA methodology defines resources in the foreground, middle ground, and background, but more specifically refers to these as level 1: internal aesthetics; level 2: relational aesthetics; and level 3: environmental aesthetics. The process begins with a thorough understanding of the characteristics of the proposed project. The visual environment is then described in terms of landscape components in the region and in the immediate project area, including land form, water, vegetation, and artificial development.

The methodology uses landscape units, either spatially enclosed or unenclosed, to define visual character. Major view sheds are defined and can be mapped. Visual resources within the view sheds are described both for roadway users (view from the road) and for roadway neighbors (view of the road). Viewers are characterized as to possible sensitivity.

Use of the FHWA technique results in numeric values for the visual character of the existing environment in four visual pattern elements (form, line, color, and texture) and four pattern characters (dominance, scale, diversity, and continuity) for each of the four landscape components (land form, water form, vegetation, and development). Visual quality also is assigned numeric values by averaging the three criteria of vividness, intactness, and unity. Each criterion is evaluated in terms of several components and at each level (1, 2, and 3). The resultant tables of visual quality values serve as the baseline, or before-project, conditions upon which to compare the values obtained by calculating after-project conditions. The difference is the degree of visual resource impact.

13.3 Impact Analysis

The impact on visual resources is defined by the changes produced by the proposed project or action. Before-project and after-project visual characters should be compared, keeping in mind that change is usually, but not always, adverse. In natural areas, such as national forests or parks, change from a natural visual environment would be adverse, based on users' expectations. For example, if the view shed from a trail is changed from forested mountain slopes and streams to a coal mine or condominium complex, the change is considered adverse. If a proposed project or action includes regrading and replanting a previously disturbed area, or removing vegetation to permit viewing of attractive vistas, the changes are beneficial.

In an urban area, more emphasis may be placed on the compatibility with the architectural style and mass of the built setting including line, form, texture, and color. Construction of a modern, straight-lined, glass-and-steel tall building within a neighboring area of low-rise, Victorian, or art deco architectural buildings is certainly considered incompatible. A change in an urban area also may be beneficial, based on the urban resident viewers' expectations. For example, if an urban blighted area is replaced by new structures, plantings of vegetation, and attractive streetscape elements, the change constitutes a positive visual impact.

Another visual ingredient, particularly in an urban setting, but possibly also applicable to some rural environments, is the changes that may be produced in shade and shadow, or light, and glare. A proposed project or action that introduces shade, shadow, light, or glare may be considered visually intrusive and incompatible with before-project conditions.

The analyst should describe the changes as objectively as possible, without drawing personal conclusions concerning attractiveness and keeping in mind that not all viewers have the same values or sensitivity to visual landscape. To a bridge engineer, an arching structure of steel or concrete over a chasm can be truly beautiful. To the trout fisher on the stream in the chasm, the same response may not be forthcoming.

In the same example, it is important to realize that individual design elements can be used to lessen the impact on the visual environment. Perhaps a massive, bulky, straight bridge design totally incompatible with the existing setting was dismissed from consideration in favor of a design that is more compatible. The bridge can be designed to be as harmonious as possible in line, form, color, and texture with its surroundings and thus be as least visually obtrusive as possible.

Computer imaging can be extremely useful in presenting before and after photographs in the environmental document. Technology exists to superimpose a proposed project into a photograph of existing settings. Perspective drawings also can assist readers in visualizing the after-project visual settings. The study team should be cautioned that few persons attending a public meeting or workshop can look at engineers' or architects' drawings, or topographic mappings, and be able to realistically interpret these documents into a three-dimensional image of the visual appearance of the after-project setting.

In cases where visual quality objectives have already been designated, as with the previously described federal management programs or with local comprehensive plans, the evaluation of visual impacts will relate directly to compatibility with established goals and objectives. For example, the BLM uses a systematic contrast rating process to analyze potential impacts (U.S. Department of Interior, BLM 1986b). The degree of impact depends on the visual contrast created between the project and the existing landscape. The contrast is measured by comparing project features with the major features in the existing landscape through use of the basic design elements of form, color, line, and texture. The basis of the comparison is established by the designated VRM objectives.

Elaborate methodologies, such as that of the FHWA, should be used with caution and only if the project characteristics warrant such detailed analysis. A problem inherent in the use of numeric evaluation methodologies is that the resultant number values imply an objectivity that perhaps does not actually exist. The subjectivity in the methodology comes in the initial assignment of numeric values to various individual components in the early stages of the process. These subjective numeric values become incorporated into formulas and techniques that imply an objective result, when in fact the result may be just manipulated subjective values. Probably the most risky aspect of using these elaborate numeric methodologies is that the answer is understood by few. The public will usually relate better to a photograph and verbal description of an area's appearance than to a result of 2.5 or 4.6.

Construction-related activities, such as ground clearance and earth movement, can normally be considered to produce adverse visual effects. The environmental document should describe the amount of earth to be disturbed and the duration of particular phases of construction activities. Mitigation measures can be used to shield adjacent receptors and should be incorporated into project construction contract documents.

These are some of the major factors in the assessment of potential visual resource impacts of a proposed project or action:

- Identify and describe visual changes.
- Consider possible changes in light, glare, shade, and shadow.
- Be objectively descriptive.
- Consider foreground, middle-ground, and background vistas and view sheds.
- Assess compatibility by comparing the form, line, color, and texture with those of the existing setting.
- Compare dominant features before and after project implementation in consistency of scale or mass.

- Make use of computer imaging, photographs, and perspective drawings.
- Consider viewer sensitivity in evaluating the degree of impact.

The content of the environmental document related to visual impacts may be a few paragraphs or several pages, depending on the identified appropriate scope of study. If extremely detailed studies are warranted, data and analysis details should be contained within a technical report and summarized within the Environmental Assessment and FONSI or Draft and Final Environmental Impact Statement.

13.4   Cumulative and Secondary Impacts

Assessment of possible cumulative impacts should consider whether the proposed project or action will incrementally add adverse visual impacts in a particular area or view shed to a degree to cause a significant overall impact. Another type of cumulative impact may include precedent-setting or policy decision types of projects. By permitting one small project or action with minimal visual impact, is pressure then created for a continuation of additional such projects or actions that cumulatively would produce adverse effects on sensitive visual resources?

Secondary impacts also may occur in terms of geographic separation or time separation. An example would be a project requiring fill material. The area from which the fill earth is to be removed may not be near the project, but would create adverse visual effects on surrounding receptors. An action permitting particular resource recovery may, in turn, promote the development of processing facilities for that resource.

13.5   Mitigation

The use of measures to reduce potential visual impacts is varied and often very project-specific. Visual buffers can obscure unsightly construction areas or land uses, such as automotive parts suppliers (junkyards). Design elements and architectural treatments can repeat basic elements of form, line, color, and texture to reduce contrast and ensure compatibility in urban areas. Mine areas should be reclaimed into natural contours and vegetation. Limiting the clearance of land for timbering can blend edges and make cleared areas less obtrusive. These examples are given to represent just a few of the many available techniques. It is important in designing mitigation measures that the degree and severity of impact be properly identified and that suitable mitigation techniques be developed with a goal of reducing impacts to acceptable levels.

14

Air Quality

Because air pollution can directly cause health risks to the human population, improvement of the U.S. air quality has been an important goal of environmental legislation and regulations since 1970. This chapter generally describes the basics of an air quality study. It is not meant to be a "cookbook" of methodologies, but rather to give the reader an understanding of the coordination and consultation requirements and the steps included in most air quality analyses. The goal of this chapter is to provide an ability to understand the purpose and results of an air quality study and to effectively communicate with specialists who conduct the actual analysis.

14.1 Legislation and Terminology

The Clean Air Act of 1970, as amended in 1990, establishes a mandate to reduce emissions of specific pollutants via uniform federal standards. Under the act, the Environmental Protection Agency is responsible for standard-setting programs and for approval of individual actions such as state implementation plans (SIPs) and specific permits.

14.1.1 Standards

Standard-setting programs under the act include the National Ambient Air Quality Standards (NAAQS), New Source Performance Standards (NSPS), National Emission Standards for Hazardous Air Pollutants (NESHAP), Acid Rain Program, Stratospheric Ozone Protection Program, and Standards for Control of Air Pollution from Motor Vehicles.

The National Ambient Air Quality Standards set maximum allow-able ambient concentrations for the following criteria air pollutants: ozone (O_3); nitrogen dioxide (NO_2); sulfur dioxide (SO_2); carbon monoxide (CO); particulate matter with aerodynamic diameter less than or equal to 10 micrometers (PM_{10}); and lead (Pb). Of these, ozone is not actually emitted directly from any source, but is formed in the presence of sunlight from two *precursor* pollutants that are emitted—oxides of nitrogen (NO_x) and reactive organic gases (ROGs), which are predominantly hydrocarbons (HCs). Primary standards are given to indicate levels necessary to protect public health; secondary standards are given for levels to protect public welfare from any known or anticipated adverse effect of a pollutant. The act requires attainment of the primary standards within a target date. Each state must attain the secondary standards within a "reasonable time." The National Ambient Air Quality Standards are shown in Fig. 14.1. These standards, and approved measurement techniques, are subject to revision

Criteria pollutant	Federal primary* standard	Sources
O_3 (1-hour average)	120 ppb	Vehicles, solvent use
CO (1-hour) (8-hour)	9 ppm 35 ppm	Vehicles
NO_2 (annual average)	53 ppb	Fuel combustion, as in vehicle engines, power plants, refineries
SO_2 (annual average) (24-hour)	30 ppb 140 ppb	Fossil fuel combustion, chemical plants, metal processing, power plants, boilers, petroleum processing
PM_{10} (24-hour) (annual mean)	150 $\mu g/m^3$ 50 $\mu g/m^3$	Dust, industrial and agricultural operations, combustion processes
Pb (calendar quarter)	1.5 $\mu g/m^3$	Previously, fuels and gasolines
TSP (24-hour)	260 $\mu g/m^3$	Fumes, dust, industrial plants

*Secondary standards are the same except for SO_2, which has a secondary standard of 500 ppb for a 3-hour average.

ppm = parts per million; ppb = parts per billion; $\mu g/m^3$ = micrograms per cubic meter

SOURCES: Los Angeles County Metropolitan Transportation Authority 1993; County Sanitation Districts of Los Angeles County 1994.

Figure 14.1 National ambient air quality primary standards and major sources of criteria pollutants.

over time. The analyst should use the most recent and applicable standards at the time of project assessment.

In addition to the National Ambient Air Quality Standards, many states have enacted air quality legislation with specific state standards. In some states, such as California, the state standards are more stringent than the national standards, and more pollutants have been added.

14.1.2 Nonattainment and state implementation plans

Geographic areas that do not meet the primary NAAQS are termed *nonattainment*. Each state has been required to develop a plan to implement the Clean Air Act, called the state implementation plan (SIP), and to demonstrate how attainment of the primary NAAQS will be reached within a target date. Air quality regulatory jurisdiction normally falls within designated smaller geographic areas, called *air basins*, and managed by air quality management districts or air pollution control districts. These agencies must prepare air quality attainment plans containing specific measures to reduce criteria pollutant concentrations below federal and state standards.

The 1990 amendments defined five classes of increasing nonattainment: marginal, moderate, serious, severe, and extreme. The only area in the United States classified as extreme for any particular pollutant is the Southern California Air Quality Basin; the pollutant is ozone, the primary ingredient of smog.

The purpose of the state implementation plan is to eliminate or reduce the severity and number of violations of the NAAQS and to achieve expeditious attainment of the standards. Federal activities may not cause or contribute to new violations of air quality standards, exacerbate existing violations, or interfere with timely attainment or required interim emission reductions toward attainment.

Geographic regions previously designated nonattainment pursuant to the Clean Air Act Amendments of 1990 and subsequently redesignated to attainment are called *maintenance areas* and are subject to the requirement to develop a maintenance plan.

14.1.3 Conformity

New and modified projects proposed within nonattainment and maintenance areas must be reviewed to determine conformity with the SIP. For stationary sources, any new emissions for a criteria pollutant or its precursors in a nonattainment area cannot be permitted without elimination of an equal or greater amount of the same pollutant through *offsets*. The offset required usually increases with the distance between the proposed and eliminated sources, but is never less than 1.0, to ensure that no net increase will occur.

The SIP outlines requirements for specific kinds of pollutants in each area of a state. The specific requirements are then applied to specific sources (e.g., individual factories or plants) by incorporating them into operating permits for each source. The Clean Air Act gives to the Environmental Protection Agency the review and approval responsibility for state implementation plans and for new-source construction permits. The EPA also establishes New Source Performance Standards and National Emission Standards for Hazardous Air Pollutants.

For transportation projects, EPA regulations require the Department of Transportation (DOT) and the Metropolitan Planning Organization (MPO) to determine conformity with the SIP on regional metropolitan transportation plans before they are adopted. Highway and transit projects funded or approved by the Federal Highway Administration (FHWA) or the Federal Transit Administration must be found to conform before they are approved or funded by the DOT or the MPO. These regulations apply to transportation-related pollutants (ozone, carbon monoxide, nitrogen dioxide, and particulates) within areas designated nonattainment or subject to maintenance plans under the Clean Air Act. The EPA also establishes motor vehicle emission standards.

The required analysis to demonstrate conformity can be quite complex and varies from state to state. Acceptable criteria for analysis methodologies and for concluding a project's conformance can be modified over time.

14.1.4 PSD review

Within areas where the NAAQS are not violated, the Prevention of Significant Deterioration (PSD) program is aimed at maintaining air quality better than the NAAQS by controlling emissions from stationary sources. Individual operating permits under PSD and new-source review (NSR) must be reviewed and approved by the Environmental Protection Agency. A PSD review for a proposed action is not required if emissions of each attainment pollutant from stationary sources would be less than a particular threshold rate.

14.1.5 Section 309 review

Another significant component of the Clean Air Act is that it establishes, within Section 309, the authority of the Environmental Protection Agency to review and rate environmental legislation, regulations, and documents for all proposed federal actions. The EPA Office of Federal Activities and its ten regional administrators rate Draft Environmental Impact Statements based on a set of criteria and provide recommendations to the lead agency for improving the draft. EPA's criteria for Section 309 review of environmental documents are shown in Fig. 14.2. If improvements are not made in the Final Environmental Impact Statement, the EPA may refer the Final Environmental Impact Statement to the Council on Environmental Quality. The council may issue findings and recommendations, may determine that the issue is not a matter of national importance, or may uphold the EPA's position. The EPA may identify a major federal action significantly affecting the environment even though the lead agency disagrees.

Rating Environmental Impacts

LO   Lack of objections

EC   Environmental concerns—impacts identified that should be avoided. Mitigation measures may be required.

EO   Environmental objections—significant impacts identified. Corrective measures may require substantial changes to the proposed action or consideration of another alternative, including any that was either previously unaddressed or eliminated from the study.

EU   Environmentally unsatisfactory—impacts identified are so severe that the action must not proceed as proposed. If these deficiencies are not corrected in the final EIS, EPA may refer the EIS to CEQ.

Rating Adequacy of the Impact Statement

1   Adequate—no further information is required for review.

2   Insufficient information—either more information is needed for review, or other alternatives should be evaluated. The identified additional information or analysis should be included in the final EIS.

3   Inadequate—seriously lacking in information or analysis to address potentially significant environmental impacts. The Draft EIS does not meet NEPA and/or Section 309 requirements. If not revised or supplemented and provided again as a Draft EIS for public comment, EPA may refer the EIS to CEQ.

SOURCE: Adapted from U.S. Environmental Protection Agency 1995c.

Figure 14.2 EPA's criteria for Section 309 review of impact statements.

14.2   Determining Existing Air Quality

To effectively predict future air quality impacts of a proposed project or action, it is necessary to first describe the air resource characteristics within the region and study area. The description should include information on climatic conditions and on pollutant levels, both current and historic. Sources of pollutants exceeding the NAAQS should be explained, if possible.

14.2.1   Existing measured data

Climatic information includes temperature, precipitation, and typical wind velocity and direction. Sometimes a *wind rose diagram,* which graphically indicates the percentage of time that annual wind occurs at particular speeds and directions, is presented within the environmental document. Vertical mixing and dilution information includes the normal mixing height above ground level in summer. These inversion layer heights are normally given for the morning and the afternoon, and they reveal important information on whether specific emissions, such as of NO_x and ROGs, are being dispersed or remain trapped. The project area or region should be reviewed for any topographical constraints affecting air pollutant dispersion characteristics.

Regional air quality management districts, or control districts, are responsible for monitoring and reporting air pollutant characteristics within their area or region. The nearest established air quality monitoring stations should be identified and presented on a map. Data from monitoring stations may need to be adjusted for the particular study area to reflect background air pollutant levels, after accounting for nearby sources of pollution.

Another source of air quality data can be previous monitoring programs associated with permit applications or environmental documents for other projects within the same, or comparable, region or area. For projects located in rural areas, assumed values sometimes can be used if they are conservative enough to be accepted by all interested and review agencies and the public.

If appropriate data is not available, an air quality monitoring program may be undertaken to provide recent, site-specific data on the existing air quality. A monitoring program must last for several months, often during specific months of the year determined to have the worst-case air pollution conditions. Therefore, a monitoring program should be undertaken only when the specific issues and scope of the proposed project or action warrant such intense effort, or when agreement on background levels of air pollutant using available data cannot be reached through coordination with the local air quality district, state air quality agency, and the EPA.

14.2.2 Modeled existing air quality

Computer models are used in the prediction of air pollutant concentrations at specific project sites. Required input for the models normally consists of "background" pollutant levels. *Background* means existing ambient air quality in areas relatively free of pollutants. Often background levels are measured in remote areas to ensure a true background not influenced by particular pollutant sources. When monitoring stations cannot be located in remote areas, resulting data may sometimes need to be adjusted (reduced) to eliminate nearby pollutant contributions. Because of this characteristic of background ambient air quality, the analysis often will compute existing pollutant concentrations in the specific project area by adding the local pollutant sources to the background concentrations.

The presentation of two different *existing* air pollutant concentrations can be confusing to the public. There are usually a *measured* level of particular pollutants and a *modeled* level for the specific site characteristics. The modeled concentrations reflect the application of the computer air pollutant dispersion model, which uses the background concentrations, but then adds site-specific pollutant sources. An example would be a busy highway intersection or a large manufacturing plant. The measured background pollutant concentrations would ideally not be monitored in such locations, because there is obviously the presence of local pollutant sources. Therefore, the background air quality would represent the intersection or factory sites with no cars or manufacturing. To get a true indication of existing pollutant concentrations, the traffic and factory emissions would be predicted with the computer model, resulting in a modeled existing description of air quality.

14.2.3 Sensitive receptors

Another activity in describing the existing air quality environment is the identification of sensitive receptors. Sensitive receptors may be residential areas, schools, parks, hospitals, or other sites for which there is a reasonable expectation of continuous human exposure during the period coinciding with peak pollutant concentrations. Air quality modeling sites for prediction of future-year pollutant concentrations are located at representative sensitive receptors in the project area. Sensitive receptors should be shown on a map in the environmental document.

14.3   Predicting Air Quality Impacts

The air quality impact of any particular proposed project or action is the difference in future-year pollutant concentrations between a no-build alternative and the proposed project or action alternatives.

If conducted productively, the scoping process may limit or focus either the pollutants to be evaluated or the level of required analysis for air quality studies for a particular proposed project or action.

14.3.1 Methodology report

For proposed projects or actions requiring detailed air quality analysis, preparation of an air quality methodology report can ensure agreement on assumptions, appropriate models, and other input factors prior to the expenditure of time and effort on the actual analysis and computer modeling. Here is an example of the many specific items which should receive agreement prior to analysis:

- Background pollutant concentrations
- Wind speed and direction
- Atmospheric stability factor
- Dispersion characteristics and mixing heights
- Sensitive receptors and modeling sites
- Traffic input and assumptions
- Emission factors
- Model to be used
- Appropriate future analysis year
- Future conditions and other projects to be incorporated (can be increased development projects, or perhaps a decrease, such as a military base closure)
- Performance of various mitigation measures
- Control measures already contained within the state implementation plan or attainment plan

The proposed assumptions and methodologies should be described in detail, and the report should receive review and concurrence (in writing) from jurisdictional and interested agencies, such as the EPA, state air quality control boards, the metropolitan planning organization, and local planning officials.

14.3.2 Graphical solutions and dispersion models

For relatively simple projects, graphs can be used to predict future regional, and sometimes local, pollutant levels. Graphical calculations have been developed for use by incorporating typical emission factors of specific types of pollutant sources and general project and site features.

Results are often obtained in tons per year or pounds per year. This type of *burden analysis* is normally used at a regional level. Screening procedures also can be used at the project level. The procedure, for a transportation project, would use worst-case assumptions and consider project location, nearby receptors, traffic volumes and level of service, and air quality conditions for current and future analysis years.

For more detailed microscale studies, air pollutant dispersion models are used to predict future pollutant concentrations with and without the proposed project or action. Input to the models is extensive and can include many features of the proposed project, area topography, distance to sensitive receptors, atmospheric and climatic data. The appropriate emission factors to be used can be based on research literature, data from similar facilities, test data on specific equipment, or, as is the case with transportation projects, mobile vehicle emission factors developed through calculations of a multitude of contributing features. With vehicles, such features would include the mix of car and truck traffic, the specific pollutant output of vehicles, the incorporation of pollutant-reducing engine features in new vehicles and the phasing out of old vehicles over time, and the existence of a mandatory state or local inspection and maintenance program for vehicles as a means to check for acceptable emission performance of the vehicle before issuing a registration.

The computer models used in predicting air pollutant concentrations vary by project type and characteristics and will not be discussed in detail in this text. Specific programs exist for transportation projects, and normally they predict carbon monoxide levels. Carbon monoxide (CO) is used as an indicator for transportation projects. Results are given in 1-hour and 8-hour parts per million (ppm) concentrations to coincide with the NAAQS for carbon monoxide. Increased CO levels are associated with congested traffic conditions and with any situation where cars are stationary and idling. Cold starts in geographic areas with cold winter weather also increases CO levels.

Results of the air quality analysis will be predicted future-year concentrations of pollutants at sensitive receptor sites for both the no-build alternative and the various action alternatives. Impact adversity is determined by comparing future concentrations with the appropriate standards, and by discussing any increases over the values of no-action conditions.

The air quality analysis also should consider related impacts that may be distant from the proposed physical site of the project. For example, changes in traffic patterns due to a new facility may cause reductions of air pollutant emissions in other areas or on other transportation routes by diverting traffic to the new facility. However, location of a new major employer or construction of a particular transportation facility may cause motorists to use residential streets as shortcuts.

14.3.3 Hot-spot analysis

In addition to predicted CO concentrations along the proposed transportation corridor, pollutant concentrations must be considered for key intersections. Separate models and methodologies and air quality criteria (normally for CO and PM_{10}) exist for these hot-spot analyses. The quantitative particulates analysis is especially applicable to bus terminals, transfer points, and commuter rail terminals, which increase the number of diesel vehicles congregating at a single location.

As noted in Chap. 9 on traffic and transportation, obtaining reliable traffic volume and flow characteristics for existing and for future build and no-build conditions in the project area is the absolute basis for obtaining reliable predictions of future CO and other transportation-related pollutant concentrations. For example, quantitative modeling hot-spot analysis is required for all intersections operating at the level of service D, E, or F or that will change to D, E, or F due to increased traffic volumes related to the proposed new project. The analyst also must model the top three intersections based on highest traffic volume and the top three based on worst level of service.

Particularly with transportation projects, for example, a new freeway or improved freeway access ramps, there may be an improvement in localized impacts at other locations. For example, if the existing freeway was so congested that motorists exited and sought alternate routes through residential neighborhoods or commercial arterial streets and the proposed improvement provides adequate capacity to permit efficient noncongested traffic flow, then the air quality impact on the previously used alternate routes will be improved.

14.3.4 Stationary sources

For other types of projects, such as factories or power plants, stationary-source computer models are used. The analysis begins with an identification of project features that would be sources of air pollution. Emission characteristics or factors are found in the literature or are developed specifically to suit the circumstances of the proposed project or action. Emissions are estimated from emission factors and characteristics of the emission device (for example, height, diameter, gas velocity) to create input information needed for the dispersion model, which is then used to compute air quality concentrations. Examples of currently used models are the EPA's Industrial Source Complex (ISC2) dispersion model and the updated industrial source complex short-term (ISCST3) dispersion model. These models compute concentrations at specific receptor points for criteria pollutants and for toxic air pollutants. Toxic air pollutant concentrations are used to compute carcinogenic, chronic, and acute health risks.

Criteria pollutants include nitrogen oxides (NO_x), reactive organic gases (ROGs), particulates (PM_{10}), oxides of sulfur (SO_x), and carbon monoxide (CO). As noted previously, ozone (O_3) is not directly emitted from mobile or stationary sources, but is formed in the presence of sunlight from precursors of nitrogen oxides and ROGs. Depending on the wind characteristics and atmospheric mixing conditions, NO_x and ROGs can be trapped at low levels in the atmosphere during nighttime and then transformed to ozone during daylight hours.

As with NO_x and ROGs, emission of sulfur dioxide can lead to other air quality impacts. Sulfate aerosols, which can originate by conversion of sulfur dioxide, account for a large percentage of visibility impairment in the desert southwest of the United States (EPA 1995). Sulfur dioxide emissions also contribute to the formation of acid rain and other sources of acidic deposition that can threaten wildlife and vegetation.

Models for calculation of toxic pollutants are based on assessing the risks due to sources of hazardous air pollutants. Toxic compounds can potentially cause three types of health risk: carcinogenic, chronic, and acute. Both carcinogenic and chronic risks are long-term and are based on the annual average ambient air quality concentrations, while acute risk is short-term and is based on 1-hour average concentrations. A carcinogenic health risk is assumed significant if the probability of toxics causing excess cancer over a lifetime at a receptor site where people reside exceeds 1 in 100,000. A chronic or acute risk is assumed significant if the hazard index for either type of risk exceeds 1.0 at a receptor site where people reside.

14.4   Cumulative and Secondary Impacts

Regional management plans help ensure that cumulative air quality effects are considered. All projects must be shown to conform with the state implementation plan and the attainment plan. As noted previously, proposed projects or actions within attainment areas must undergo EPA review under the Prevention of Significant Deterioration (PSD) or New-Source Review (NSR) procedures. The PSD procedures establish increments of prescribed levels of air quality degradation that will be allowed in an area. New projects are not permitted to violate these PSD increments. The various legislative and regulatory standards, management plans, attainment plans, and permit requirements ensure that no particular project or action will violate federal or state standards, contribute to an existing or projected violation of the standards, exacerbate the violation of a standard, contribute to a delay in attainment of standards, or be inconsistent with an approved air quality attainment plan.

The analysis of cumulative impacts must, of course, be sure to include other proposed projects within the region or area. Such inclusion is ensured, for example, with the requirement that all proposed transportation projects be included in the regional transportation improvement program at the time the entire program is analyzed for conformity with the state implementation plan. If a particular project was not included when the latest analysis was completed, the entire system must be rerun with the incorporation of the proposed project to ensure that cumulative effects with other proposed projects are considered.

Potential secondary air pollution impacts must be investigated based on the characteristics of the particular project to ensure that no possible effects are overlooked. For example, a proposed landfill project must consider the air quality impact of transport of the wastes to the site, even to the point of assessing the possible impacts of vehicles waiting longer periods at particular at-grade rail crossings, and employee trips to and from the facility. There would also be emissions of construction vehicles, particularly since a landfill type of operation basically requires continuous construction activities.

Secondary air quality impacts also may occur when the proposed project or action allows specific resources to be harvested, which in turn will be processed at other, sometimes distant, locations. This type of secondary impact, for example, was a particular issue for an EPA Environmental Impact Statement on a proposed permit for strip coal mining activities in Texas, near the Mexican border. The operator of the mine intended to sell the coal for use at a power plant complex about 20 miles away in Mexico. The power plant, although in Mexico, is a major source of existing and future air pollution emissions which impact a national park and other receptors on the U.S. side of the border. Substantial technical, policy, and public controversy existed over the power plant, especially since the coal provided by the mine in Texas had a lower heat value than that of many other coals which could be burned. Therefore, use of the coal from the Texas mine would require a greater amount to be burned to supply the same heat output for electricity production and would produce a greater amount of sulfur dioxide. A secondary secondary impact was that much of the power generated at the plant is in turn used at a steel mill not far from the power plant. This example illustrates how consideration of secondary and cumulative impacts is not always easy or straightforward.

Quantification of possible secondary impacts also is very difficult for sulfur dioxide emissions. Sulfur dioxide may convert to sulfate aerosols, which in turn can cause severe impairment of visibility. Reduced visibility, in turn, can produce other types of secondary impacts. Sulfur dioxide in the atmosphere can cause acid rain and other sources of acidic deposition. The resultant effects on water quality, vegetation, and wildlife are indeed secondary impacts of the emission of sulfur dioxide, but are extremely difficult to quantify or predict.

Regarding stationary sources, the emission of criteria pollutants by a particular source may consume much of the PSD increment designated in the state implementation plan. The air pollutant emissions therefore could restrict future industrial growth, which in turn would affect employment and income to the region.

14.5 Mitigation

Mitigation of air quality impacts is not a new or recent area of study. Since the passage of the Clean Air Act in 1970, and even before that time, successful methods for reducing harmful emissions into the atmosphere have been analyzed and developed. Results of these studies have produced numerous techniques, equipment, and measures for reducing potential air pollutant concentrations to within the NAAQS and to systematically improve air quality over time. This section cannot describe all available methods, but will give examples of some types of measures in use.

14.5.1 Mobile-source pollutants

In many areas of the United States, transportation vehicles are the major cause of violations of the NAAQS. State implementation plans and air quality attainment plans contain methods to reduce mobile-source emissions to reach attainment within target dates. These transportation control measures can vary from very technical equipment changes, such as requiring engines to emit lower concentrations of particular pollutants, to public behavioral changes through education and encouragement to use carpools and mass transit systems. These are examples of some typical transportation control or management measures:

- Mandated vehicle emission reductions by particular target years
- Use of clean fuels
- Conversion of bus fleets from diesel to electric
- Use of natural gas vehicles
- Construction of high-occupancy vehicle (HOV) lanes
- Construction of mass transit systems
- Appropriate spacing of freeway entrance and exit ramps to prevent congestion

- Ramp meters during peak hours to regulate the flow of traffic
- Improvements to intersection signalization and traffic flow characteristics
- Incentives at the workplace to carpool or use mass transit
- Installation of message signs on freeways to inform of congested areas
- Provision of adequate shoulder areas for disabled vehicles
- Implementation of call boxes and a full-time towing service to rapidly remove disabled vehicles
- Staggered work hours and flexible time for major employers
- Reduction in locomotive emissions or replacement of diesel engines with electrified systems
- Requirement of all proposed transportation projects to be included in the regional transportation plan found to conform with the state implementation plan

The analysis of possible mitigation measures for a particular proposed project or action must be sure not to take credit for any measures already committed within an adopted attainment plan. The goal is to produce a net reduction of pollutant levels by implementing additional measures for the specific proposed project.

14.5.2 Stationary sources

Mitigation measures for stationary-source pollutants are very specific to the type of project. A regional landfill may incorporate liners to prevent the escape of landfill gas. A power plant will include scrubbers to remove sulfur oxides and nitrogen oxides. Mitigation techniques are therefore very industry-oriented and should be developed for the specific characteristics and problem emissions of the proposed project or action.

In some circumstances, emissions of nonattainment pollutants and precursors can be mitigated by *offsets* of equal or greater reductions of each emitted pollutant, such that no net increase occurs. The offset rate can vary; an example would be a need to offset 1.3 times the emission rate. The proposed offset activity need not be within the jurisdiction of the project sponsor. An example is a proposed landfill project in rural California which proposes offsets related to a diversion of agricultural plant material from burning. The elimination of agricultural burning within the county, which is primarily a source of PM_{10}, ROGs, NO_x, and CO, was shown to offset the increases of nonattainment pollutants attributable to the landfill.

14.5.3 Mitigation assurance

In addition to avoiding double counting, or taking credit for measures already committed within an attainment plan analysis for future years, proposed project-specific mitigation (emission reductions) can be claimed only to the extent that implementation is ensured. Evidence should be provided for the commitment to implement. Such evidence should include

- An identification of responsible parties
- A schedule for implementation
- Evidence of availability of funding
- Procedures for monitoring and enforcing

Only if this evidence of assurance of implementation of proposed mitigation measures is established and in place at the time of the environmental analysis can credit emission reduction be taken.

14.6 Contents of Environmental Document

The existing air quality environment should be described in the environmental document through the use of tables and graphics, with text to describe special characteristics or reasons for particular air pollution events. Tables present available data on criteria pollutants relevant to the characteristics of the proposed project or action and compare the concentrations with the NAAQS and any state air quality standards. Particularly, the discussion should focus on any nonattainment pollutants and possible sources.

Historic air quality trends can be compiled, showing number of days per year that a particular pollutant exceeded the standards and the highest concentration reached. A sample graphic of the type that may be used is shown in Fig. 14.3. Explanation of the violations of the standards should be given if possible, such as wind direction or other atmospheric conditions, particularly if the pollutant source is not in the project area. Often geographic areas of nonattainment are due not to pollution generation within that area, but to transport of pollutants from other, more populated areas.

The environmental document should describe the air quality planning for the region and site. The plans contain target dates for attainment and specific control measures to be implemented to meet standards. Coordination with state, regional, and local air quality management agencies is absolutely essential throughout the air quality study, including input on methodologies to be used for calculating existing and future air pollutant levels.

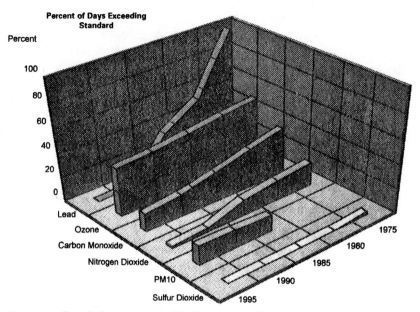

Figure 14.3 Sample historic air quality data presentation.

The preparer of the Environmental Assessment/FONSI or the Draft/Final Environmental Impact Statement must at all times remember the audience for whom the document is written, that is, the public. Presentation of results of an air quality analysis should be direct, simple, and easy to understand. Details on existing data, monitoring programs, model assumptions, and analysis of computer data should be contained in a separate technical supporting document or in the appendix of the environmental document.

The environmental document should contain tables showing future air pollutant levels with and without the proposed project or action. All proposed no-build and build alternatives should be equally assessed. The discussion should focus on which, if any, pollutant concentrations will exceed the applicable standards. Conformity with the SIP and any proposed mitigation measures should be summarized. Remember to document all consultation and coordination efforts with EPA, state air resources boards, regional metropolitan planning organizations, air quality districts, and local planning organizations. Opinions of these agencies should not be hidden or obscure in the document, even if a difference of opinion among agencies or the project sponsor occurs.

15

Noise

Although not as obviously affecting public health, noise impacts can definitely be important to a local community and often become highly contested issues at public meetings. Sensitivity to noise increases, along with visual impacts, is probably the most obvious perceived impact by the public and yet very variable and somewhat subjectively individual in how a particular person will react or be affected. These features of noise make it a challenging impact to assess and study. Standards, or impact criteria, exist, but are based on nuisance levels, reduction in ability to hear conversation, or ability of the human ear to detect differences in noise levels.

15.1 Terminology, Standards, and Criteria

Noise is defined as unwanted sound. Sound becomes noise when it interferes with normal everyday activities such as sleeping, reading, and conversation. Noise becomes a health hazard when it adversely affects hearing ability or causes psychological harm that could, in turn, lead to degradation of physical health.

15.1.1 Noise-level descriptors

Noise is expressed in decibels (dB), the basic unit for measurement of sound. Because the human ear has a different sensitivity to noise sources than a microphone, a logarithmic weighting curve, the A-weighted scale, has been developed for use in approximating the sensitivity of the average human ear perception of loudness. Therefore, noise levels related to human impacts are measured and expressed in terms of A-weighted decibels (dBA). Figure 15.1 indicates the approximate dBA noise levels for several common sounds.

To assist in the assessment of noise levels most representative for particular noise sources and environments, various agencies and municipalities have developed measurement scales, or noise descriptors,

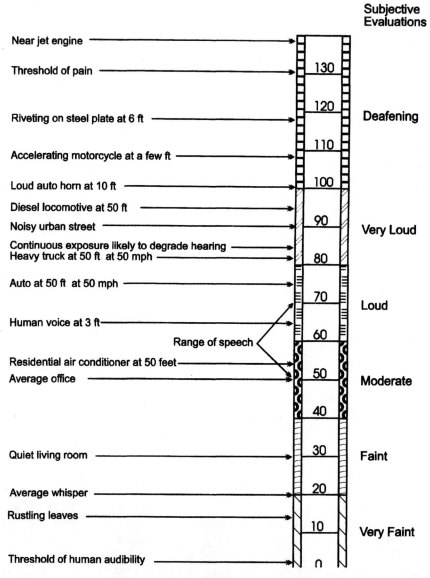

Subjective Evaluations

Near jet engine

Threshold of pain — 130

120

Riveting on steel plate at 6 ft — **Deafening**

110

Accelerating motorcycle at a few ft

Loud auto horn at 10 ft — 100

Diesel locomotive at 50 ft

Noisy urban street — 90   **Very Loud**

Continuous exposure likely to degrade hearing
Heavy truck at 50 ft at 50 mph — 80

Auto at 50 ft at 50 mph

70   **Loud**

Human voice at 3 ft

Range of speech — 60

Residential air conditioner at 50 feet
Average office — 50   **Moderate**

40

Quiet living room — 30   **Faint**

Average whisper — 20

Rustling leaves — 10   **Very Faint**

Threshold of human audibility — 0

Figure 15.1 Typical noise levels. (Adapted from UMTA and City and County of Honolulu, 1989, and FHWA, CA DOT and SBAG, 1992B.)

for averaging, calculating, and representing noise levels. Time-varying sound levels are often described in terms of an equivalent constant decibel level. Equivalent sound levels, denoted by L_{eq}, are used to develop single-value descriptions of average noise exposure over various periods of time.

For example, the Federal Highway Administration (FHWA) uses this equivalent noise level, or L_{eq}, by averaging the dBA noise levels measured over a specified time. A 1-hour L_{eq}, used by the FHWA in assessing highway noise levels, is the average of instantaneous dBA sound levels measured over 60 minutes. The actual noise level measured by sound meters during 1 hour may vary between 60 and 70 dBA, and the 1-hour L_{eq} may be calculated to be 66 dBA, depending on the percentage of time the noise level was at the high or low end of the scale. The result is a single number that can be compared to federal and state noise criteria established for highway projects.

In some descriptors, average noise exposure values include additional weighting factors for annoyance potential attributable to time of day or other considerations. Average noise exposure over a 24-hour period is often presented as a day-night average sound level, denoted by L_{dn}. The L_{dn} values are calculated from hourly L_{eq} values, with the L_{eq} values for the nighttime period (10 p.m. to 7 a.m.) increased by 10 dB to reflect the greater disturbance potential from nighttime noises.

The community noise equivalent level (CNEL) also is used to characterize average noise levels over a 24-hour period, with weighting factors included for evening and nighttime noise levels. The L_{eq} values for the evening period (7 p.m. to 10 p.m.) are increased by 5 dB, while the L_{eq} values for the nighttime period (10 p.m. to 7 a.m.) are increased by 10 dB.

Infrequently occurring noise sources may be described just in terms of the maximum noise occurrence, without any averaging. Examples may include the use of a peak pass-by noise level for trains and peak short-term duration of noise emitted from construction equipment.

15.1.2 Comparison of descriptor use values

A peak hourly L_{eq} value for traffic noise will differ from an associated L_{dn} value depending on the distribution of traffic over the 24-hour period. There is no precise way to convert a peak hourly L_{eq} value to an L_{dn} value. In urban areas exposed to high traffic volumes, the peak hourly L_{eq} value is typically 2 to 4 dBA lower than the daily L_{dn} value. The relationship will change with the degree of development to an equality in value, and, finally, to a peak hourly L_{eq} value, 3 to 4 dBA greater than the daily L_{dn} value in rural areas with little nighttime traffic.

An L_{dn} value calculated at a site from a set of measurements taken over a given 24-hour period will be slightly lower than the CNEL value calculated over the same period. Except in situations where unusually high evening noise levels occur, the CNEL value will be within 1 to 2 dBA of the L_{dn} value for the same set of noise measurements.

15.1.3 Characteristics of noise

Because noise-level scales are logarithmic, values cannot be directly added to each other to calculate a total combined noise level. Two noise sources producing equal sound levels at a given location will produce a composite sound level that is 3 dBA greater than either sound alone. When two values differ by 10 dBA, the composite noise level will be only 0.4 dBA greater than that of the louder noise source alone.

Human ear perception of change in noise level is generally accepted to be as follows:

- 3 dBA or less is barely perceptible, if at all.
- 5 dBA is clearly perceptible.
- 10 dBA is perceived as being twice (or half) as loud.

Because noise consists of sound waves traveling through air, levels decrease with the distance from the noise source. With no intervening obstruction, noise from a single source will decrease approximately 6 dBA for every doubling of distance away from the source. When the noise source is essentially a continuous line, such as vehicle traffic on a highway, noise levels generally will decrease about 3 dBA for every doubling of distance.

When intervening land or structural features are present within the distance between the noise source and the receptor, noise values can be affected by these features. If intervening ground is covered with noise absorption materials, such as grass, shrubs, and trees, the reduction in noise levels will be somewhat greater than the 3-dBA value noted above for traffic noise. Topographic features and structural barriers can cause sound waves to be absorbed or to bounce and reflect in different directions, thereby affecting the value of noise at a particular receptor. Atmospheric conditions also can affect the degree to which sound is reduced over distance.

15.1.4 Guidelines and criteria*

Based on characteristics of noise sources and receptors, several federal agencies have developed guidelines for evaluating land-use compatibility for applicable noise-level ranges. The federal Noise Control Act of 1972 established a requirement that all federal agencies develop programs to promote an environment free of noise that jeopardizes public health or welfare. Although the Environmental Protection Agency has responsibility under the act, each federal agency has the authority to adopt noise regulations pertaining to that particular agency's activities. For example, the Occupational Safety and Health Administration sets workplace noise exposure standards, and the Federal Aviation Administration sets aircraft noise standards.

Under the federal Noise Control Act, the Environmental Protection Agency identified indoor and outdoor noise limits to protect public health and welfare (communication disruption, sleep disturbance, and hearing damage). The Federal Highway Administration has established noise criteria for evaluating impacts associated with federally funded transportation projects on the basis of justification of funding noise mitigation measures. The FHWA criteria are based on two assessments: absolute noise values and the increase over existing levels regardless of resultant absolute level. If the project causes noise levels exceeding the noise abatement criteria, then mitigation measures are incorporated into the project design. The Department of Housing and Urban Development has established guidelines for evaluating noise impacts on residential projects seeking financial support under various grant programs. Summaries of the basics of these agency standards and criteria are given in Fig. 15.2 as examples of variations in guidance for different types of proposed projects or actions.

In other cases, noise criteria may be specifically developed based on the particular source or proposed project. The Federal Transit Administration uses a noise exposure (L_{eq}) criterion consisting of a maximum acceptable increase over existing levels. Criteria also have been developed specifically for transit and other types of periodically recurring noise, using a maximum pass-by descriptor L_{max}. Even more specific criteria have been developed, such as design goal criteria for substation noise or for vehicle noise. Examples of these industry-specific criteria are shown in Fig. 15.3.

An important element of the federal Noise Control Act is that it directs all federal agencies to comply with applicable federal, state, in-

---

*Although this section is derived from several sources, the information on state and local examples relied predominantly on County Sanitation Districts of Los Angeles County, 1994.

Figure 15.2 Examples of federal agency noise criteria.

Environmental Protection Agency: Noise Limits to Protect Public Health and Welfare

Protection against:	Outdoor limit	Indoor limit
Speech interference and sleep disturbance for residential, educational, and health care areas	55-dB L_{dn}	45-dB L_{dn}
Hearing damage in commercial and industrial areas	70-dB 24-h L_{eq}	70-dB 24-h L_{eq}

Federal Highway Administration: Exterior Noise Abatement Criteria

Activity category	Maximum 1-h L_{eq}
A: Land where serenity and quiet are of extraordinary importance	57 dBA
B: Picnic areas, playgrounds, residences, motels, schools, churches, libraries, and hospitals	67 dBA (52 dBA indoors)
C: Developed lands	72 dBA

Department of Housing and Urban Development: Guidelines for Evaluating Noise Impacts on Residential Properties

Acceptability for residential use	Outdoor L_{dn} levels
Acceptable	65 dB or less
Normally acceptable	65–75 dB
Unacceptable	Greater than 75 dB

SOURCES: County Sanitation Districts of Los Angeles County 1994; U.S. Department of Transportation, FHWA, CADOT, and SBAG 1992b.

terstate, and local noise control regulations. Many states have guidelines and standards for evaluation of noise impacts and the requirement to incorporate mitigation measures into the proposed project or action. For example, the California Department of Health Services published guidelines for the noise element of local general plans that include a noise-level versus land-use compatibility chart. The chart categorizes various outdoor L_{dn} ranges into differing degrees of acceptability based on land use. The California Department of Housing and Community Development has adopted noise insulation performance standards for new hotels, motels, and dwellings other than detached single-family structures. The standards set requirements for maximum indoor noise levels, from outdoor sources, for any habitable room with windows closed.

Figure 15.3   Examples of industry-specific noise criteria: transit noise.

Criteria for Maximum Noise from Train Operations

Community area category	Maximum pass-by noise levels, dBA		
	Single-family dwellings	Multifamily dwellings	Commercial buildings
Low-density residential	70	70	75
Average residential	70	70	80
High-density residential	70	75	80
Commercial	75	80	85
Industrial	80	80	85

Criteria for Maximum Noise from Train Operations near Specific Types of Buildings

Building or occupancy type	Maximum pass-by noise level, dBA
Amphitheaters	65
"Quiet" outdoor recreation areas	70
Concert halls, TV studios	70
Churches, hospitals, schools, theaters, museums, libraries	70–75

Guidelines for Noise from Substations

Community area category	Maximum noise level design criteria, dBA
Low-density residential	35
Average residential	40
High-density residential	45
Commercial	50
Industrial	60

SOURCE: U.S. Department of Transportation, UMTA, and City and County of Honolulu, 1989.

Municipalities also establish local noise guidelines, usually within the adopted general plan and through noise ordinances. The county of Los Angeles and the city of Los Angeles are used here as examples. The county of Los Angeles General Plan Noise Element establishes noise-related goals and policies and describes the general noise environment. The county also has a noise ordinance which recommends maximum expected ambient noise levels for four land-use categories: noise-sensitive, residential, commercial, and industrial areas. If, however, a measured ambient noise level at a specific project location exceeds the expected levels outlined in the ordinance, the measured level may be used as the baseline noise level.

The county's exterior noise standards are described in terms of maximum time duration for exceeding the baseline at particular increases in noise levels. For example, the baseline noise level plus 10 dBA may not be exceeded for more than 5 minutes in any 1-hour period. Interior noise standards set maximum allowable noise levels and set time duration for any violation of these standards by specific increases over the allowable levels. The county also has separate restrictions for construction-related noise, based on time of day and maximum levels for various land uses. The construction standards also distinguish between short-term and long-term construction projects.

The city of Los Angeles General Plan Noise Element gives a value for use in assuming the expected ambient noise level in residential areas during the day. The city's noise ordinance states that this expected ambient noise level in a given area may not be increased by more than 5 dBA. If actual measured noise levels are above the stated expected ambient level, then the measured levels may be used as the baseline for determining whether there is a 5-dBA increase. The L_{eq} noise descriptor is used for analysis, and the 5-dBA increase criterion applies to both noise generated by construction activities and long-term operation-generated noise.

The discussions and examples in this section should make clear to the environmental analyst the importance of researching the applicable noise standards and criteria for the project or action being proposed.

15.2   Methodology Report

As with air quality studies, it is often advantageous for projects requiring detailed noise assessments to prepare a separate noise analysis methodology report. The report should be used as a vehicle for coordination and concurrence on selected sensitive receptors; proposed measurement program; assumptions; applicable federal, state, and local standards or criteria for impacts assessment; and the particular computer model program to be used for predicting future noise levels at sensitive receptor sites.

If an adverse noise impact is expected, the methodology report should discuss available mitigation techniques and their acceptability for use. Agreement should be obtained on the performance success, or achieved attenuation, for specific measures. For example, all interested parties should reach agreement that a particular noise barrier or wall will provide a particular noise reduction at particular heights, lengths, and composition of materials.

15.3   Describing the Existing
Noise Environment

After consultation with appropriate officials and agencies to establish applicable standards and criteria, the expected impact area of the proposed project or action should be delineated. Sensitive receptors, such as schools, churches, parks, residential neighborhoods, hospitals, and libraries within the expected impact area, should be identified. Noise measurements with industry-approved equipment are then taken at selected representative sensitive receptor sites. The number of measurements, duration of measurement, and time of day will be dictated by the particular methodology selected as appropriate based on characteristics of the proposed project or action. During actual measurements, notations are made of any unusual occurrences which may distort a true representation of ambient noise levels, such as dogs barking, airplanes overhead, or excessive vehicle horns. Based on the selected descriptor, the measured sound levels are then converted to L_{eq}, L_{dn}, CNEL, or peak pass-by values.

The Environmental Assessment or Draft Environmental Impact Statement should contain a map of sensitive receptor locations, a list of what the sensitive receptors are, and a table of existing noise measurements at each measurement site. A brief discussion of the existing noise environment should describe the general characteristics, any major contributing sources, or other features important for the public to understand the general nature of the ambient noise environment throughout the day.

For areas near transportation facilities, existing noise levels sometimes can be determined without actual monitoring or measurement. The FHWA has noise prediction models which estimate average noise levels at fixed distances from the roadway centerline based on roadway geometrics, estimated traffic volumes for automobiles and medium- and heavy-duty trucks, vehicle speeds, and a designated noise drop-off rate. Shielding effects from topographical features, buildings, and other barriers are not accounted for in the model; it thus produces a conservative, worst-case estimate of traffic-generated noise levels.

The description of the existing noise environment in the environmental document may contain two values for existing noise levels: measured and modeled. The environmental analysis must consider and compare future noise levels of the no-action alternative with future noise levels of proposed build, or action, alternatives. Therefore, future existing noise levels are calculated with a computer model. For transportation projects, the measured noise levels and simultaneously collected traffic counts are used to calibrate the noise model. The calibration compares the expected noise levels with the actual measured noise levels and thus permits correction to account for effects of topography and building shielding unique to a particular sensitive receptor site. If the analysis does use both measured and modeled values for existing and future existing noise levels, the reasons for the two values and the value for use as a baseline for determining future noise increases should be described in the environmental document in a straightforward, easy-to-comprehend manner through the use of tables or graphs.

15.4 Predicting Noise Impacts

As with most areas of impact analysis, the first step in noise impact assessment is to reach a complete understanding of the specific characteristics, physical and operational, of the project or action being proposed. For transportation projects, such information will include physical parameters of the actual facility and traffic volume and flow characteristics. For stationary sources or construction activities, information required will include type of equipment, hours of operation, and specific noise-producing attributes of various equipment.

As with air quality studies, future noise levels at particular sensitive receptor sites are predicted by using a computer model. There are some major differences in important considerations between air quality and noise studies. For example, whereas the worst-case air quality impact will normally occur during peak-hour traffic conditions with maximum congestion, noise levels may actually be quite low during peak traffic hours. Because a major source of noise from vehicles is generated by tire movement on pavement, vehicles moving at slow speeds during congested conditions do not produce much noise compared with vehicles moving at higher speeds. The worst-case noise peak hour is thus normally considered to be when traffic is operating at level of service C, or moving at maximum volumes with little delay.

For this reason, future noise-level calculations often are dependent not on actual projected traffic volumes, but rather on the level of service C volume for the particular physical design features of the highway. Particularly for highway projects on a new location, data input to the computer model relies on physical characteristics of the proposed highway (width, grade, etc.), details of projected traffic characteristics (volume, speed, percentage truck mix), and topographic or other buffering or intervening land features.

Use of level of service C volumes is based on the assumption that at some period during the day, the highway does actually operate at level of service C. For projects that propose widening or other alteration of an existing highway facility, it must be considered that the existing facility also most likely operates at level of service C at some time during the day. The evaluation of impact may therefore rely mostly on changes in physical characteristics and perhaps the duration of level of service C conditions.

Another consideration that becomes more obviously important in noise studies as compared with air quality studies is the incorporation of information on the possible removal of structures and other changes in land form that may affect sensitive receptors protected from noise impacts in before-project settings. For example, a row of houses immediately adjacent to an existing freeway may be destroyed as the freeway is widened. Not only is traffic physically closer to the next row of houses due to widening of the freeway, but also the buffer of the row of houses that previously shielded those second-row homes has now been removed.

Stationary sources are normally assessed for noise impacts by using industry data and literature review on the actual noise produced by the employed equipment or noise source. Evaluation of construction-related noise also relies upon known information on the noise produced by various equipment and activities at individual stages of construction. For example, noise levels produced at 50 ft are about 84 to 85 dBA from backhoes and bulldozers, 91 to 92 dBA from graders, 80 to 88 dBA from compressors, 85 to 98 dBA from jackhammers, and 96 to 107 dBA from pile drivers (U.S. Department of Transportation, FHWA, CADOT, and SBAG 1993; County Sanitation Districts of Los Angeles County 1994). Review of the construction schedule and associated duration of each type of noise-producing activity is used to determine the degree of noise impact on sensitive receptors. The analysis includes assumptions on noise levels from the equipment, the drop-off rate for distance, an atmospheric absorption coefficient, and distance to the receptor. Effects of shielding due to intervening ground topography or structures also may be considered.

In some cases, use of more than one descriptor may be appropriate to fully explore possible noise impacts. Analysis methodologies may need to include multiple modeling approaches. For example, projects that produce intermittent loud noises, such as trains, airplanes, and perhaps certain types of manufacturing or resource recovery activities, may not be appropriately assessed by using an L_{eq} descriptor or L_{eq} standards and criteria. The L_{eq} method would incorporate the intermittent loud noise occurrences into an hourly average and thus may not truly represent the level of disturbance to nearby sensitive receptors. In these types of circumstances, the noise impact analyst should consider presenting both the L_{eq} descriptor and perhaps results of a peak occurrence analysis to determine effect.

The particular conditions of a proposed project may require a rather detailed analysis of effect of intervening features or physical parameters. For example, a noise source below or above a sensitive receptor would not produce the same noise levels as if the source were at the same elevation. The presence of structures or walls may redirect noise to receptors otherwise not expected to be affected.

Results of a noise analysis will be computed future dBA levels at selected sensitive receptor sites. These noise values are then compared with existing measured or modeled values, and with applicable standards and criteria. The comparison should be between the various proposed action alternatives and the no-action alternative. Locations where the appropriate standards and criteria are exceeded require consideration of mitigation.

15.5 Ground-Borne Vibration and Noise

Certain types of projects have the potential to produce vibration waves through the ground as well as through air. Trains are the most common source of vibration impact, so rail projects will be used here as examples of criteria, impact assessment, and mitigation.

Ground-borne vibration and ground-borne noise are the same phenomenon up to the point of perception at the dwelling or building. Ground-borne vibration describes waves in the ground which can be measured by using vibration equipment mounted on sidewalks, foundations, or stakes in the ground, and which can be perceived as mechanical motion. Ground-borne noise describes sound generated when the same waves in the ground reach room surfaces in buildings, causing them to vibrate and radiate sound waves into the room.

Vibration surveys result in data in velocity levels, to compare with established criteria. Although studies about vibration and associated impacts are not extensive, it is generally thought that weighted vibration levels below 69 dB are generally imperceptible, or just perceptible, to the average person under normal conditions (U.S. DOT, UMTA, and City and County of Honolulu 1989). Criteria have been established for maximum pass-by ground-borne vibration velocity levels for train operations for various land uses and for particular types of buildings. Similar criteria exist for ground-borne noise, measured in A-weighted decibels.

Because the magnitude of vibration transmitted from the source to receptor buildings depends largely on soil conditions within the intervening geologic strata, the analysis of potential impact is normally based on very site-specific characteristics. Specific characteristics of the proposed project and the receptor buildings also are extremely significant factors in the impact assessment. For example, a steel wheel and steel rail train will produce substantially more vibration than a rubber tire type of light-rail transit system. Specific significant differences can result from the type of track, type of vehicle, use of a rotary versus a linear induction motor, and location at ground level, above ground level, or below ground.

Another example may be the specific location of operational equipment, such as frog switches for train crossovers. A steel wheel and steel rail train crossing the gap inherent in switches will generate significantly higher vibration levels than operation on standard continuous trackwork. It is therefore necessary to locate such switches as far as possible from vibration-sensitive buildings, such as residences or hospitals, and preferably in industrial or undeveloped areas.

Many times, exact determination of possible ground-borne vibration and noise impacts will not be made during the environmental analysis phase of the project. The reasoning for this approach is that the level of detailed study required could not feasibly be conducted for all locations of all possible considered alternatives. Therefore the detailed analysis of impacts and required mitigation is normally reserved until the preliminary engineering phase of the selected alternative.

There would be no reason to even consider vibration impact at the environmental study phase if results could not contribute to the decision-making process. These studies do become useful, however, by making a more general analysis of potential impact during the environmental impact assessment phase. Using train operations again as an example, the comparative evaluation of alternatives can be developed based on expected distance from the rail, or base of column for aerial systems, beyond which vibration impacts would likely not occur.

For example, a study may conclude that vibration levels from a steel wheel and steel rail system would be acceptable for residential structures beyond 35 to 40 ft, while the vibration levels from a beam-straddling monorail system would be acceptable beyond 15 to 20 ft and would be lower than or equal to those due to trucks and buses on local roads. This type of information, combined with variations among the alternatives in location alignment relative to location of sensitive buildings, can then permit an impact comparison among proposed considered alternatives.

15.6 Cumulative and Secondary Impacts

The assessment of cumulative impacts should consider other proposed projects within the same impact area. By use of existing measurements, existing noise sources will be included. Proposed future projects or actions, however, may need to be incorporated into the evaluation of cumulative impacts.

Secondary impacts are most easily identified by thoroughly reviewing all the ramifications of the proposed project or action. For example, if a factory will use raw material to be supplied by trucks or rail, the evaluation of secondary impacts should include the sensitive receptors along the truck haul routes and the rail line. If a proposed project is a major employer, secondary impacts may occur due to increased traffic on local streets accessing the new facility. Construction equipment needs to be transported to and from a project site, and often soil or other materials will need to be hauled away. Again, the assessment of possible secondary impacts should include proposed haul routes. Detours of vehicular traffic will similarly expose receptors along detour routes to increased noise levels.

15.7 Mitigation

Mitigation of potentially adverse noise impacts can be accomplished through a variety of methods. Noise-producing equipment can be modified, such as with mufflers, to produce less noise. Railroad and transit noise can be reduced through changes in wheel or track design. Noise barriers, or walls, are often used to shield sensitive receptors from highway noise. If there is sufficient space, earthen barriers may also be considered. At times, the project sponsors will make changes to the receptor rather than to the noise source, such as installing air conditioning so that windows may remain closed at all times. Restricting time of day for noise-producing activities also can be used to ensure a minimized difference in noise levels.

Noise barrier design can be quite complicated and is again based on the use of computer models to determine the exact attenuation of noise levels obtained with varying wall heights, lengths, and materials.

It is acceptable for a project-sponsoring agency to set feasibility criteria for implementation of mitigation measures. For example, the criteria for determining whether to construct a noise wall to shield sensitive receptors from highway noise may include factors such as maximum attainable reduction in noise, number of residents or homes to benefit, and cost of the wall per benefited receptor. In some cases, it may be more feasible and cost-effective to purchase the home(s) and relocate the residents than to construct a noise barrier.

All proposed mitigation measures, such as walls, that visibly or otherwise directly affect a neighborhood, church, school, etc., should be developed with productive input from the affected receptor. Some neighborhoods, for example, may rather have the noise than a wall.

Ground-borne vibration and noise impacts mitigation is very industry- and site-specific. Examples of types of successful mitigation techniques for rail systems would include use of special frog switches, ballast mats or floating slab under the track, special bearing pads between the track structure and support column, soft and resilient rail fasteners, or avoiding a rigid connection with adjacent buildings.

15.8   Contents of Environmental Document

If a detailed noise impact analysis is conducted, a separate supporting technical report should be prepared. The technical report contains detailed information on assumptions, methodologies, measurement equipment and survey, model calibrations, and results of impact and abatement analyses.

The Environmental Assessment/FONSI or Draft/Final Environmental Impact Statement should contain summary tables and discussions required to understand the results of the analysis and lead to a better decision on selection of an alternative. Maps should show locations of sensitive receptors, measurement sites, modeling sites, and noise barriers, if proposed. Tables should indicate measured and modeled future existing noise levels and the calculated noise levels for each proposed action alternative. Tables should indicate the resultant noise levels before and after proposed mitigation implementation.

As with other areas of impact assessment, all coordination with regional and local agencies, and relevant community opinions derived through the public participation program, should be summarized and addressed.

16

Geology and Soils

The assessment of potential impacts related to geology and soils is significantly interrelated to other areas of impact assessment, particularly water quality. Although it is difficult to separate a discussion of groundwater from geology, all discussion of water quality and quantity takes place in Chap. 18 on water resources. Geology and soils effects may differ from those of other disciplinary areas of assessment because many proposed projects or actions will not actually cause effects *on* the geology or soils of a site or an area. Effects, rather, are normally *associated with* geology or soils as opposed to causing any physical or chemical changes in the characteristics of the actual geology or soils. There are exceptions, however, such as contamination of soils (discussed in Chap. 17 concerning hazardous wastes) or loss of productive soil.

16.1 Describing Existing Resources

The level of detail on geology and soils required to determine the potential for impact, and thus included in Section III of the DEIS or within the EA, will greatly depend on the type of project alternatives being proposed. Projects requiring excavation or tunneling or causing changes in drainage patterns may dictate more detailed studies than those with no substantial surface disturbance.

16.1.1 Geologic features

As a basis for impact assessment, the analyst should obtain an understanding of the basic physiography and topography of the site or area and the underlying geology. Local planning documents and discussions with local agencies will normally provide geologic information.

The U.S. Geological Survey (USGS) also is an excellent source of information on geology, geologic hazards, groundwater quality, and surface water flow characteristics. The USGS is the principal source of scientific and technical expertise in the earth sciences within the federal government and has been providing data and reports for more than 100 years (U.S. Department of Interior, USGS 1995f).

Information and mapping will indicate important characteristics of geologic formations, such as stability for construction, permeability and porosity, groundwater aquifers, seismicity and faults, sinkholes, springs, natural gas or oil wells, surface and deep mines, mineral resources, and volcanic activity. Engineering limitations and constraints may include difficulty of excavation, cut slope stability, and foundation stability. In earthquake-prone areas, regional risk assessment maps are available from the USGS. Based on collected ground-motion data and other geologic information, the hazard maps provide estimates of the probability of significant ground movement and the potential areas of landslides, mud flows, and liquefaction. Liquefaction is the temporary change of a saturated soil or fill to a liquid, which produces a corresponding loss of support strength for structures.

16.1.2 Seismicity

If applicable to the study area, existing information on faulting and seismicity should be included in the environmental document. Often information on faults will be available on a regional basis and on a very localized basis through local universities or independent experts. Previous environmental studies for other projects in the area are a valuable source of information. Normally, major faults will be readily identified, but minor faults may not be identified or may be identified only as "potential."

Detailed studies on existing faulting and seismicity characteristics would be conducted by qualified geologists or geotechnical engineers. In some cases, it may be necessary to excavate test pits or trenches to investigate fault potential. The trenches would permit determination of the age of the subsurface material and evidence of surface rupture through examination of whether underlying strata are continuous or discontinuous. Discontinuous strata could indicate fault ruptures.

Information available from a seismicity study will normally include the maximum credible earthquake (MCE) and the maximum probable earthquake (MPE) magnitudes. The MCE is the largest possible earthquake considering the known tectonic framework of an individual fault, and the MPE is the largest earthquake likely to occur with a 100-year return period. Associated with these estimated maximum earthquake magnitudes, the peak horizontal ground acceleration is calculated. The duration of strong ground shaking for each fault also may be estimated.

The seismicity information in the environmental document should include history of earthquakes in the project area or region, their magnitudes, and the locations of epicenters. Characteristics of the earthquake activity, as related to study area faults, should be described, if known.

16.1.3   Mineral resources

The description of mineral resources should include a summary of potential mineral ores, natural gas, oil, geothermal resources, and sand and gravel. Mineral ownership, existing and proposed resource recovery activities, and the potential for resource development should be described, if relevant to the characteristics of the proposed project or action. The potential for undiscovered mineral resources has already been assessed for some geographic areas, and it depends mostly on the underlying geology. For example, an area's potential for oil and gas pools in a favorable spatial relationship is rated high if three geologic necessities exist: (1) a geologic trap, an impermeable lithologic barrier, to prevent the oil and gas from escaping to the surface; (2) suitable reservoir rock, a unit with sufficient porosity and permeability to hold a quantity of oil and gas and transmit it when penetrated by drilling; and (3) mature source rock, usually a carbon-rich shale or limestone, which could have generated hydrocarbons during burial, compaction, and heating. Most of the lands within national forests or under the jurisdiction of the Bureau of Land Management have been mapped for low, moderate, or high potential for each of various mineral resources.

After the potential for resource is known, nongeologic factors influence whether the resources are actually feasible to develop. These economic-setting factors include the market value of petroleum and the geography of the area. Geography is important in calculating the cost of recovery, such as rugged terrain, distance from support facilities, and distance from markets, which may make the expected returns from drilling insufficient to cover extra expenses incurred.

Both the potential presence of mineral resources and economic feasibility enter into the determination of potential for resource development. The potential for development should be included within the environmental document if it is predetermined, mapped, and relevant to the type of proposed project or action being considered.

16.1.4   Soil surveys

Information and mapping on soils are available in soil surveys from the Natural Resources Conservation Service (NRCS), formerly the Soil Conservation Service (SCS). For each soil map unit, the soil survey provides an abundant amount of information related to other areas of potential impact. Depth to groundwater, erodibility, and drainage characteristics will be important to assess the potential impact on water resources. Depth to bedrock, permeability, available moisture capacity, and suitability or limitations for construction use may affect engineering studies and whether local soils would need to be removed and/or replaced prior to construction. Limitations identified for each soil series would include such information as high frost heave potential, possible sinkholes, flooding potential, slow permeability, or seasonal high-water table.

Information for each soil type includes the types of soils, such as silt loam, clay, and alluvium and the percentage of slope of the land within the mapping unit. The soils survey also provides interpretations for different uses, such as cropland, forest land, rangeland, homesites, recreation, wildlife habitat, and septic tank filter fields. The survey will identify prime farmland soils, soil of statewide importance, and hydric soils. Hydric soils are soils that are saturated, flooded, or ponded long enough during the growing season to develop anaerobic (no-oxygen) conditions in the upper part. Information on the presence of hydric soils is necessary for the identification of wetlands and is discussed further in Chap. 20.

16.1.5 Contents of the environmental document

Because of the abundant information on geologic resources and soils that will be available for review, it is extremely important that the environmental analyst remain focused on relevant issues and concerns. The characteristics of the proposed project or action and the results of agency and public scoping should direct the review of available information to that which may assist in impact evaluation. Special geologic constraints, hazards, mineral resources, and soil characteristics within the actual area of potential impact should be summarized in the environmental document. The presence of hydric or prime farmland soils should be indicated and mapped, if significant. The goal is to identify any possible serious constraints or conflicts in compatibility of the proposed project or action with existing geologic and soil resources characteristics.

16.2 Geologic Impacts

The evaluation of geologic impacts and geology-related impacts will be specific to the type of project proposed for implementation. Often impacts are avoided through incorporation of specific design criteria within permits or contractors' specifications. Mineral resource recovery has direct geologic impacts. Geology-related impacts can occur, particularly to groundwater and surface water resources.

16.2.1 Geologic hazards

In the assessment of possible geology-related impacts, it is important to first attempt to avoid geologic hazards or existing resources, such as sinkholes, caves, gas wells, surface mines, or gravel quarries. Emphasis again is placed on the continuing process of refining alternatives throughout the environmental impact assessment process. Alternatives should minimize impact related to geologic hazard areas as much as possible through shifts in site locations or design characteristics.

16.2.2 Land-use compatibility

In many situations, the evaluation of geology and soils-related impacts is basically an evaluation of land-use compatibility. The characteristics of the proposed project or action should be overlaid upon the critical relevant geologic and soils data to determine the compatibility of the proposed use with the existing features of the geology and soils.

For many types of projects, this evaluation of land-use compatibility has already been done and is reflected in various building codes, legislation, regulations, and permits. For example, a certain type of landfill must not be built within 200 ft of land that has been ruptured by Holocene faults (active within the past 10,000 years). Some design criteria for buildings, landfills, dams, and highway structures are based on assumptions related to seismic activity. The structures must be designed to withstand a certain earthquake magnitude and horizontal ground acceleration without damage to foundations or structures that control pollution and without risk to human life or property.

16.2.3 Mineral resources

Potential impacts on mineral resources and resource development will be most critical with environmental studies when the proposed project or action is a land-use or management plan, such as those for national forests or land under the jurisdiction of the Bureau of Land Management. The analyst again should be aware that various federal and state agencies have been established with specific directives on the resources under their jurisdictions. See the discussion in Chap. 5 on the conflicts that can sometimes arise because of differing mandates, in that example between the Environmental Protection Agency's responsibility to protect the nation's air quality and the Department of Transportation's responsibility to provide the nation with safe and efficient transportation.

As an example related to mineral exploration, and development on federally owned lands, it is frequently unknown by the general public that the same act (Organic Act of 1897) that created the National Forest System also opened the national forests to mineral development. Forest Service mineral management policy was first set in 1907. Under the caption "To the Public" Gifford Pinchot, first chief of the Forest Service, gave the following directive:

> The timber, water, pasture, minerals and other resources of the National Forests are for the use of the people. They may be obtained under reasonable conditions without delay. Legitimate improvements and business enterprises are encouraged.

Numerous directives and laws since that time have reinforced the general mineral resource management policy of the systematic discovery and characterization of mineral and energy resources so that the most important deposits can be developed and utilized to best meet the needs of society. Because the Forest Service does not discover or develop energy or hardrock mineral resources, the achievement of the policy is defined by the amount of access provided to industry so it may discover and develop the nation's mineral resources. The guideline against which mineral resource management is measured is that at least 75 percent of the federal mineral estate is available for reasonable and prudent energy or mineral exploration (U.S. Department of Agriculture, Forest Service 1992a). The area involved, however, is small; less than 1 percent of national forest land has ever been disturbed by mineral development.

Protection from adverse impact to the environment by mineral resource development is ensured through requirements for environmental impact assessment of proposed activities and for reclamation of lands disturbed by mineral and energy activities for other productive uses. Mineral exploration and development is a temporary and widely scattered land use. Mineral commodities are open to development only under leases, permits, or licenses issued by the Bureau of Land Management, which has the responsibility for leasing all onshore federal minerals. Environmental analysis is required for each stage of mineral development activities.

The direct impact of mineral resource development on mineral resources is depletion of the resource and associated benefits to the nation's energy and minerals supply and security. Indirect effects may be the creation of jobs and economic benefits at the local, state, and federal levels. Direct effects on soils may include disturbance within highly erodible soils (refer to discussion of soil erosion later in this chapter).

Mineral resource development can produce direct and indirect, secondary effects on other resources and, in this sense, is evaluated in an Environmental Assessment/Finding of No Significant Impact or Draft and Final Environmental Impact Statements, the same as any other proposed project or action. Impacts are possible to air quality, water, wetlands, recreation, visual quality, cultural resources, vegetation and wildlife, and socioeconomics.

16.2.4   Geology-related effects

Other types of impacts associated with geologic conditions include these:

- Removal and disposal of unsuitable material
- Leaching of pollutants into groundwater systems
- Interception of the water table through excavation and resultant required pumping (dewatering) during construction
- Exposure of acid-producing geologic formations to rainwater

Geologic considerations are directly related to the quantity and quality of groundwater resources. These potential groundwater and surface water impacts are discussed in greater detail in Chap. 18.

16.2.5   Interdisciplinary approach

If the proposed action is a construction project, such as a building, mall, highway, bridge, or harbor improvement, engineers will be an important part of the multidisciplinary team working on the development of project alternatives. In these cases, engineering investigations of geotechnical conditions will be conducted. These investigations will normally focus on the engineering suitability of geology and soils in the area, as opposed to natural qualities such as groundwater leaching or exposure of acid-producing formations.

There will, however, be overlap in many areas of information valuable to both engineers and environmental impact analysts. The importance of continued team interaction and sharing of information cannot be overemphasized. Often those in environmental disciplines can make engineers aware of potential problems early in the design process. Moreover, engineers will be able to directly answer questions required for the environmental team members to fully understand exactly what will occur during construction so that the degree of potential impact can be properly evaluated.

Detailed subsurface engineering studies are undertaken after the environmental impact assessment process, during final design, and before construction of the selected alternative, to ensure no major impact related to foundation or stability conditions.

16.3   Erosion

Although many people in the United States no longer think about it in any detail, the nation's soil is an extremely important natural resource for food production and public welfare. As early as 1935, the

Soil Conservation and Domestic Allotment Act recognized that "the wastage of soil and moisture resources on farm, grazing, and forest lands of the Nation, resulting from soil erosion, is a menace to the national welfare..." and established the Soil Conservation Service (now the Natural Resources Conservation Service) with the purpose of providing permanently for the control and prevention of soil erosion.

As with hydric and prime farmland soils, the Natural Resources Conservation Service has identified highly erodible soils. The agency prepares, and makes available to the public, lists of highly erodible soil map units. The determination has been made through application of the highly erodible lands criteria (7 CFR Part 12). The criteria use two basic formulas for determining the erosion rate: that due to rainfall and that due to wind. These erosion rates are then divided by a predetermined soil loss tolerance value.

The soil erosion rate due to rainfall is calculated by using the universal soil loss equation (USLE). The USLE is a multiplication of three factors: rainfall and runoff R, the degree to which the soil resists water erosion K, and a factor (LS) describing the effects of slope length (L) and steepness (S). The resulting number represents the potential average annual rate of sheet and rill erosion due to rainfall.

The potential average annual rate of wind erosion is estimated by using the wind erosion equation (WEQ), which multiplies two factors: the climatic characterization of wind speed and surface soil moisture C by the degree to which soil resists wind erosion I.

Values for all the factors used in the soil loss equations are calculated, or are already contained in the soil survey information by soil map unit, using methodologies explained in the U.S. Department of Agriculture handbooks or Natural Resources Conservation Service field office technical guides and references.

The criterion for highly erodible lands, then, is the result of the rainfall or wind erosion rate calculation divided by a factor T representing a predetermined soil loss tolerance. The T value represents the maximum annual rate of soil erosion that could occur without causing a decline in long-term productivity. The designation of highly erodible lands, therefore, is not based solely on actual erosion, but on the relationship of erosion rates to the maintenance of desired productivity, or use, of the land.

The evaluation of potential soil erosion impacts will focus on the amount of ground to be cleared at any one time, the slope of ground, erodibility of exposed soils, and rainfall potential. The universal soil loss equation can be used to predict the amount of soil potentially lost to erosion, if the project warrants a detailed assessment. Soil erosion can produce a direct impact on aquatic life in surface waters through sedimentation, as discussed in the chapter on water resources.

Fortunately, most environmental analyses of potential erosion effects for small, site-specific projects will be able to conclude that, with implementation of proper mitigation measures, the remaining impact will be negligible. The potential for impact, however, is important to establish the required mitigation. State-of-the-art erosion and sedimentation control techniques are available to all contractors and are usually contained in contract specifications in detail, including timing and staging of placement of erosion control measures related to construction activities. Measures may include such activities as seeding of bare slopes; provision of diversion drainage ditches; use of hay bales or straw around catch basins and drainage structures to detain soil; installation of fabric silt fencing around the area to be cleared; construction of detention basins; or limiting the amount of permitted bare ground at any one time. Guidelines and specifications for revegetation and erosion control practices for soil-disturbing activities can be developed on a soil-type-specific basis.

The environmental impact assessment for more comprehensive projects and actions, such as long-range management plans for extensive acres of national forests or Bureau of Land Management lands, may not be so easily concluded that successful mitigation of soil erosion impacts is possible. Plans that will set policy and specific requirements for many years regarding the use of land for grazing, timbering, and mining may have serious effects on soil conservation and productivity. Lands that are mismanaged because of incorrect assumptions or unforeseeable factors can cause irretrievable and irreversible loss of soil resources and associated watershed or stream impacts. For this reason, long-term and geographically extensive management plans must be carefully assessed. Periodic checks and balances must be included to permit reassessments and revisions based on changing environmental factors.

16.4 Soil Suitability

As with geologic features, the evaluation of soils-related impacts often involves assessing the suitability of soils to support the proposed activity or project, or suitability of the proposed project or action given the soil characteristics of the location. For example, if soils are not appropriate for supporting structures, or maintaining a slope without slippage, then unsuitable soils may have to be excavated and replaced with other soils, which are then compacted to produce required support characteristics. On the other hand, if existing ground is composed of fill material, it may not be a suitable location for the construction of homes or buildings. The goal of the analyst is to determine the compatibility of the proposed use of land with the characteristics of the soils.

The Natural Resources Conservation Service land-use policy objectives, as set forth in Parts 400 to 404 of the *General Manual,* are as follows:

- Systematically protect agricultural land, including cropland, rangeland, and forest land, from unnecessary and irreversible conversion to nonagricultural uses.

- Discourage incompatible uses of land or the construction of unnecessary encroachments in floodplains and wetlands.

- Promote the use of land within its capabilities to protect natural resources and to ensure public health, safety, and welfare.

- Encourage and assist states and local governments in planning for growth and development in coastal areas to protect the coastal resource base.

As noted in the land-use policies, soil considerations overlap with other areas of impact assessment, such as floodplains, wetlands, or coastal zones, discussed in other chapters of this text.

Legislation and regulations continue to strive for increased compatibility in land use with natural soil and geologic conditions. An example is the Watershed Protection and Flood Prevention Act to provide assistance for persons living in small watersheds and to provide additional treatment and protection of federally owned lands within such watersheds. The Food Security Act of 1985 discontinues certain benefits and incentives provided by the Department of Agriculture to persons who produce agricultural commodities on highly erodible land or converted wetland.

Although mostly considered as a compatibility assessment, some types of projects affect the chemical or physical characteristics of the soil. Examples are poor agricultural practices, clearing of land cover that produces changes in sunlight and the moisture content of soil, or application of fertilizers or pesticides to crops. Severe soil contamination also can result from spillage of hazardous materials such as fuel oil, acid mine drainage, or leachate from landfills. These types of effects can render soils unsuitable for most uses.

16.5 Farmland

The protection of farmland is directly related to a concern for the rapid conversion of agricultural land, including cropland, forest land, and rangeland, to nonagricultural uses. The conservation of highly productive agricultural land and the maintenance of sustainable food production are goals consistent with the national welfare.

16.5.1 Sustainable agriculture

Sustainable agriculture refers to agricultural practices that, through the use of technology, provide for long-term sustainability of production, profit, environmental quality, and food safety. It is achieved through management strategies which help the producer select hybrids and varieties, soil-conserving cultural practices, soil fertility programs, and pest management programs. The goal of sustainable agriculture is to minimize adverse impacts to the immediate and off-farm environment while providing a sustained level of production and profit (7 CFR Part 407).

16.5.2 Legislation and definitions

The Farmland Protection Policy Act of 1981, as amended in 1987 (FPPA), has a purpose of minimizing the extent to which federal programs contribute to the unnecessary and irreversible conversion of farmland to nonagricultural uses. The act defines three levels of important farmland: prime farmland, unique farmland, and farmland of statewide or local importance.

Prime farmland, generally, is land that has the best combination of physical and chemical characteristics for production of agricultural crops with minimum input of fuel, fertilizer, pesticides, and labor. The Natural Resources Conservation Service actually uses a number of very specific criteria to designate soil as prime farmland. Prime farmland can see current use in cropland, rangeland, or forest land, but is not already in, or committed to, urban development or water storage. Prime farmland already in urban use has a density of 30 structures per 40-acre area. Prime farmland committed to urban development must score 160 points or less from the land evaluation and site assessment criteria (discussed below).

Unique farmland is land other than prime farmland that produces specific high-value food and fiber crops. *Farmland of statewide importance* is land so designated by state agencies as important in the production of crops.

Many states also have enacted specific laws and regulations for the protection of agricultural lands. These statewide requirements can be more stringent than federal requirements, and they also would apply to non-federally assisted projects. Sometimes preparation of a separate farmlands impact assessment report, for review and approval at the state level, is required. The analyst should coordinate at the earliest possible time with state and local planners and officials to determine any special study requirements.

16.5.3 FPPA criteria

The Farmland Protection Policy Act directs the Department of Agriculture to develop criteria for identifying the effects of federal programs on the conversion of farmland to nonagricultural uses. The DOA's criteria are contained in 7 CFR Part 658 and consist of two parts: land evaluation criterion—relative value, and site assessment criteria.

The Natural Resources Conservation Service provides technical assistance to develop state and local agricultural land evaluation and site assessment (LESA) systems. Procedures for developing LESA systems are contained in the Department of Agriculture *National Agricultural LESA Handbook* (U.S. Department of Agriculture, Soil Conservation Service, 1983). The LESA system is designed to determine the quality of land for agricultural uses and to assess sites or land areas for their agricultural economic viability. The *Handbook* provides guidance on land evaluation criteria for cropland, forest land, and rangeland and on site assessment criteria.

Compliance with the Farmland Protection Policy Act is accomplished through use of Form AD-1006, the Farmland Conversion Impact Rating Form. A summary of form information is shown in Fig. 16.1.

The first part of the criteria, the land evaluation or relative value, is established by state and local officials, with technical assistance from the Natural Resources Conservation Service. Based on soil characteristics, groups of soils within a local government's jurisdiction are evaluated and assigned a score between 0 and 100, representing the relative value for agricultural production of the farmland to be converted by the proposed project or action compared to other farmland in the same local government jurisdiction. This score is the relative value rating on Form AD-1006.

The second set of criteria for evaluation of farmland conversion impact, the site assessment criteria, is used to evaluate specific site characteristics of the proposed project or action. Based on the answers to 12 questions, a maximum score of 160 points is possible for a particular site or alternative. The criteria contained in the Farmland Protection Policy Act rule include scoring a proposed site based on consideration of these 12 factors:

Criteria	*Maximum total points*
1. Area in nonurban use	15
2. Perimeter in nonurban use	10
3. Percentage of site being farmed	20
4. Protection provided by state and local governments	20
5. Distance from urban built-up area	15
6. Distance to urban support services	15

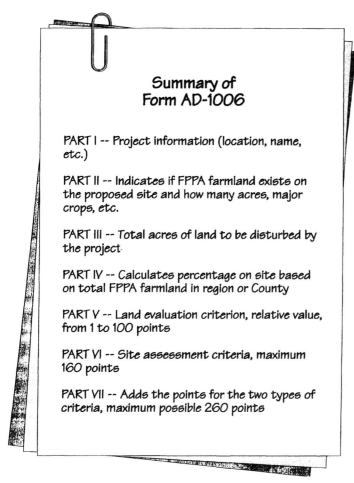

Figure 16.1 Summary of Form AD-1006.

7. Size of present farm unit compared to average	10
8. Creation of nonfarmable land	10
9. Availability of farm support services	5
10. On-farm investments	20
11. Effects of conversion on farm support services	10
12. Compatibility with existing agricultural use	10
Total possible points	160

For projects that have linear, or corridor-type sites, criteria 5 and 6 are eliminated and criteria 8 and 11 are scored on a scale of 0 to 25 points.

Although the example included above relates to the criteria contained within the Natural Resources Conservation Service guide for conformance with the Farmland Protection Policy Act, the analyst should be aware that many local areas have developed specific LESA systems for use in that particular geographic area. These local systems, if approved by the NRCS for use in compliance with the FPPA, should be used for the site assessment portion of the analysis.

16.5.4   Interpreting results

The combined land evaluation and site assessment criteria yield a maximum score of 260 points for any particular site or alternative. The highest combined score indicates sites most suitable for protection as farmland. Sites with combined scores of less than 160 points are to be given minimal protection, and no additional sites need to be evaluated. If a site receives a score of 160 points or more, the proposed project or action must be reevaluated to consider (1) the use of land that is not farmland; (2) alternative sites, locations, and designs; and (3) special siting requirements that may preclude the use of an alternative site.

16.5.5   The process

The presence or absence of soil classified as prime farmland, unique farmland, or soils of statewide or local significance is determined through review of applicable soil surveys for the sites or locations of the proposed project or action. The project sponsor fills in part I of Form AD-1006 and forwards the form, with appropriate project alternatives descriptions and mapping, to the Natural Resources Conservation Service. Refer to Fig. 16.2 for a summary of the FPPA compliance process.

After review of the information on the proposed project or action, the Natural Resources Conservation Service determines if the site is farmland subject to the Farmland Protection Policy Act. The Natural Resources Conservation Service has 45 days to respond to requests for determinations of farmlands subject to the act. If the Natural Resources Conservation Service does not respond, and further delay would interfere with construction activities, the agency may proceed as though the site were not farmland.

If farmland subject to the act is involved, the NRCS returns Form AD-1006 to the sponsoring agency, with completed parts II, IV, and V, the land evaluation sections of Form AD-1006. The sponsoring agency then applies the site assessment criteria and completes parts III, VI, and VII. Based on the total combined score for relative value (land evaluation) and site assessment criteria, the agency determines the suitability of the site for protection as farmland. A copy of the completed Form

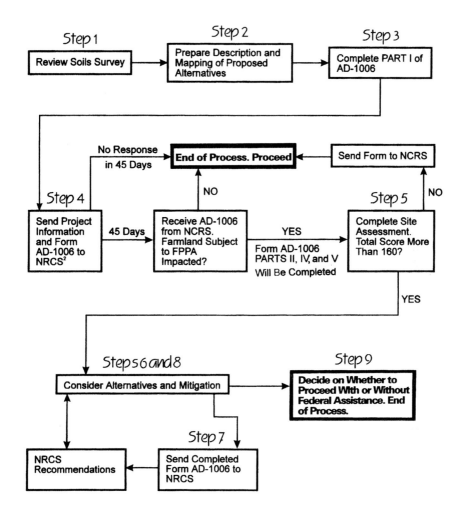

1. Farmland Protection Policy Act
2. Natural Resources Conservation Service

Figure 16.2 FPPA compliance process.

AD-1006 must be returned to the NRCS after a decision relating to farmland conversion has been made by the federal sponsoring agency.

16.5.6 Rights of private property owners

The 1987 amendments to the Farmland Protection Policy Act clarified the issue of private ownership and the discretion of the federal agency to provide assistance to convert farmland to nonagricultural uses. The act clearly states that the federal government is not authorized to regulate the use of private or nonfederal land. In a case of a private party or nonfederal unit of government applying for federal assistance to convert farmland to nonagricultural use, the federal agency will apply the criteria and recommend alternatives or measures to avoid or minimize adverse effects. If the landowners want to proceed with the proposed project or action, the federal agency may provide or deny the requested assistance. The private parties or nonfederal government unit may proceed with the proposed project or action without federal assistance.

17

Environmental Health and Public Safety

The subject of environmental health includes potential impacts related to hazardous materials, toxic substances, worker safety, and risk of upset. In the environmental impact assessment process, these studies, particularly when required to be detailed investigations, will be conducted by environmental health specialists. Interaction and coordination between members of the multidisciplinary team are essential to produce a final summary of relevant results in the environmental document that can be easily understood by the general public.

Although given a separate chapter in this text, the impacts related to environmental health are usually conveyed to the public in air quality, water, and soil resources.

17.1 Related Legislation and NEPA Requirements

The primary federal agency responsible for the control of environmental health pollutants is the Environmental Protection Agency. Several laws and regulations apply to hazardous substances and charge the Environmental Protection Agency with control responsibilities through standard setting and issuance of permits. Many of the procedures established to comply with legislation have been determined to be the functional equivalent of an environmental analysis required under the National Environmental Policy Act (NEPA). Much of this section of text has been summarized from results of an EPA work session comparing EPA programs with NEPA requirements (U.S.

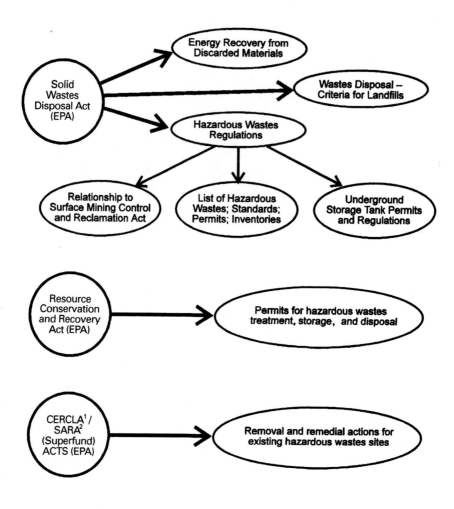

1. Comprehensive Environmental Response, Compensation, and Liability Act
2. Superfund Amendments and Reauthorization Act

Figure 17.1a Summary of environmental health and public safety laws and regulations.

Environmental Protection Agency 1993). A summary of the laws and regulations discussed in this section is shown in Fig. 17.1a and b.

17.1.1 Solid Wastes Disposal Act

The Solid Wastes Disposal Act provides technical and financial assistance for the development of management plans and facilities for the recovery of energy and other resources from discarded materials and for the safe disposal of discarded materials, and to regulate the management of hazardous waste. The act incorporates and is amend-

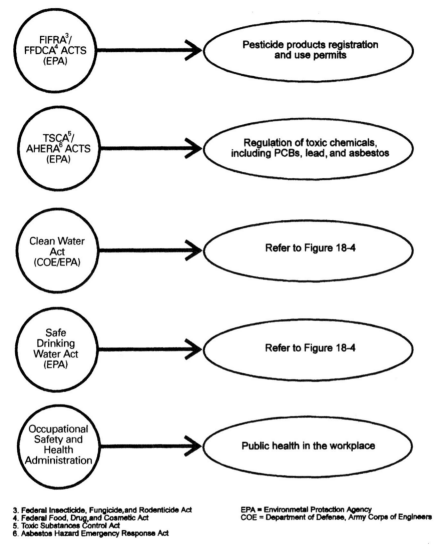

Figure 17.1b Summary of environmental health and public safety laws and regulations.

ed by other related legislation, most recently (1987) by the Hazardous and Solid Wastes Amendments of 1984, the Safe Drinking Water Act Amendments of 1986, and the Superfund Amendments and Reauthorization Act (SARA) of 1986.

The act emphasizes the needless waste of recoverable material by burying and that methods are available to separate usable material from solid wastes. By recognizing that solid wastes represent a potential source of solid fuel, oil, or gas that can be converted to energy, the act emphasizes the need to develop alternative energy sources and finds that technology exists to produce usable energy from solid waste.

The act as amended, among other regulations, establishes

- A list of hazardous wastes
- Specific standards for generators of hazardous waste, transporters of hazardous waste, and owners and operators of hazardous waste treatment, storage, and disposal facilities, including prohibiting land disposal of specified wastes
- Permit requirements
- The relationship to surface coal mining and reclamation permits issued under the Surface Mining Control and Reclamation Act of 1977 to control coal mining wastes or overburden
- Site inventories required by states, to include the location, owner, type of hazardous waste, and current status
- Monitoring, analysis, testing, and reporting requirements of owners or operators
- Regulation of underground storage tanks

The standards establish requirements for such things as record keeping; labeling; container use; a manifest system to identify the quantity, composition, origin, routing, and destination of hazardous waste from point of generation to point of disposal; treatment or storage; and reporting to state or local agencies.

Amendments to the Solid Waste Disposal Act related to underground storage tanks are predominantly derived from the Hazardous and Solid Wastes Amendments of 1984. The law requires owners of underground storage tanks to notify the appropriate state or local agency of the existence of such tanks, specifying the age, size, type, location, and uses. Each state must maintain two separate inventories of all underground storage tanks in the state, one for petroleum and one for other regulated substances.

The underground storage tank regulations include requirements for

- Maintaining a leak detection system
- Reporting releases and corrective actions taken
- Taking corrective action in response to a release
- Closure of tanks to prevent future releases
- Maintaining evidence of financial responsibility for taking corrective action and compensating third parties for bodily injury and property damage caused by accidental releases

The Hazardous and Solid Wastes Amendments of 1984 additionally required a nationwide study of underground storage tanks to include

■ Description of the tanks, including location, age, and type

■ Soil conditions, water tables, and the hydrogeology of tank locations

■ Likelihood of releases from tanks

■ Effectiveness and cost inventory systems, tank testing, and leak detection systems

The Solid Waste Disposal Act also provides federal guidelines for state or regional solid waste plans and criteria for sanitary landfills. The objectives of this part of the act are to "assist in developing and encouraging methods for the disposal of solid waste which are environmentally sound and which maximize the utilization of valuable resources including energy and materials, which are recoverable from solid waste and to encourage resource conservation."

17.1.2   RCRA

The Resource Conservation and Recovery Act (RCRA) requires the Environmental Protection Agency to set standards and issue permits to develop the regulatory framework to identify those wastes that must be managed as hazardous. Under the Resource Conservation and Recovery Act, the EPA sets requirements for management of hazardous wastes and a system for permitting facilities that treat, store, and dispose of EPA-listed or identified hazardous wastes. Standards set under the Resource Conservation and Recovery Act must be based on the protection of human health or the environment.

Hazardous waste treatment, storage, and disposal facilities are subject to permitting by the EPA. The permitting process covers the details of the design and an evaluation of environmental impact of the facility. A public notice of a draft permit decision must by published by the EPA in the local area of the facility. After a 45-day public review and comment period, the EPA makes a final permit decision. Court cases have concluded that the permitting process under the Resource Conservation and Recovery Act is the functional equivalent of the NEPA process (U.S. Environmental Protection Agency 1993).

Responsibilities under the act can be delegated to the states in lieu of the EPA when the state has a program that is consistent with, and at least as stringent as, the federal program.

17.1.3   CERCLA and SARA (Superfund)

The Superfund program is authorized under the Comprehensive Environmental Response, Compensation, and Liability Act (CERCLA) of

1980, as amended by the Superfund Amendments and Reauthorization Act of 1986. The program sets forth two types of actions to protect human health and the environment from the nation's abandoned or uncontrolled hazardous waste sites: removal actions and long-term remedial actions. Removal actions refer to short-term releases or threats of release of hazardous materials. Superfund remedial actions apply to sites on the National Priorities List (NPL), Superfund's list of highest-priority hazardous waste sites, and consist of permanently and significantly reducing the dangers, but do not necessitate removal action.

The CERCLA procedures are detailed in the National Contingency Plan and constitute a functional equivalent of NEPA compliance. These are the primary steps in the procedures:

1. *Site screening and inspection.*   The purpose is to identify hazardous substances and the sensitive populations and environmental features likely to be affected.

2. *Hazard ranking system score.*   This quantitative estimation of the relative threat to humans and/or ecosystems determines whether the site is placed on the National Priority List.

3. *Remedial investigation/feasibility study (RI/FS).*   This study determines the nature and extent of contamination and evaluates proposed alternative remedies. A detailed risk assessment, including exposure pathways and toxicity levels, is required. Proposed remedial alternatives are objectively assessed by using nine evaluation criteria (U.S. Environmental Protection Agency 1993):
   - Overall protection of human health and the environment;
   - Compliance with applicable, relevant or appropriate requirements;
   - Long-term effectiveness and permanence;
   - Reduction of toxicity, mobility, or volume through treatment;
   - Short-term effectiveness;
   - Implementability;
   - Cost;
   - State acceptance; and
   - Community acceptance

4. *Public review.*   An opportunity for review and comment by the public and interested agencies, states, or tribes is afforded. The community involvement program also includes interviews; meetings, fact sheets, press releases, workshops, and/or site tours as determined suitable for the specific site; and a technical assistance grant of $50,000 to local community groups, allowing them to hire an adviser who can both monitor and explain the highly complex, technical aspects of a typical cleanup.

5. *Record of decision.* This documents the background information on a site and describes the selected cleanup method and how it was selected. Responses to comments received during the review period are included.

6. *Site visits.* Every five years, the EPA visits sites where the selected remedy leaves contaminants at levels that do not allow for unrestricted access to ensure the remedy is still protective.

CERCLA also established the Agency for Toxic Substances and Disease Registry which conducts health assessments based on exposure through releases of hazardous wastes.

17.1.4 Pesticides

The Environmental Protection Agency regulates the conditions of distribution, sale, and use of pesticides through authorization provided by the Federal Insecticide, Fungicide, and Rodenticide Act (FIFRA) and the Federal Food, Drug, and Cosmetic Act (FFDCA). The procedures for compliance with registration of pesticide products and granting use permits are functional equivalents of an Environmental Impact Statement under NEPA. The process requires assessment of potential environmental impacts from the use of pesticide products and a public review and comment period.

17.1.5 Toxic substances

The Toxic Substances Control Act (TSCA) authorizes regulation of toxic substances including polychlorinated biphenyls (PCBs). Additional regulation of lead and asbestos is contained within the Asbestos Hazard Emergency Response Act (AHERA). TSCA provides prevention of unreasonable risks of injury to health or the environment from a chemical at any stage in its life cycle—manufacturing, processing, distribution, use, or disposal. It also provides a permitting program for disposal procedures for PCBs.

The act requires preparation of risk assessment (RA) and regulatory impact analysis (RIA) documents to evaluate environmental impacts. The process includes consideration of alternatives, including chemical substitutes, pollution prevention options, process changes, and substitute products. A public review and comment period is required. The permitting and regulatory procedure is thus a functional equivalent of NEPA.

17.1.6 Clean Water Act

The Clean Water Act contains numerous provisions for the protection of public and environmental health, some of which are summarized here.

The EPA establishes guidelines for development of water quality criteria by the states and sets national water quality criteria for specific chemical pollutants. The national water quality criteria are based on testing to provide protection for aquatic organisms and for human health. Water quality criteria programs developed by states or tribes must contain the following:

- Use designations, which require that all waters, where attainable, be designated for propagation of fish, shellfish, and wildlife and for recreation in and on the water
- Use attainability analyses for stream segments that do not meet the goal
- Water quality criteria to protect in-stream uses
- Antidegradation policy to maintain existing uses

The Section 401 water quality certification process provides an opportunity for states, tribes, and the EPA to ensure that water quality standards are met.

Regulations for water quality planning require that states identify waters that are not anticipated to attain or maintain water quality standards and develop a priority ranking. The planning process must include development of total maximum daily loads (TMDLs) that define the specific reductions in chemical and other forms of pollution necessary to protect water quality.

Section 404 of the act establishes a program to regulate the discharge of dredged and fill material into waters of the United States, including wetlands. Section 404 permits are discussed in greater detail in Chap. 18, Water Resources.

The Clean Water Act establishes the National Pollutant Discharge Elimination System (NPDES) program (also referred to as *Section 402 permits*). Effluent standards and guidelines regulate industrial discharges to surface waters and to publicly owned treatment works. The limitations and standards applicable to direct discharges are implemented in NPDES permits for point sources discharging directly to the waters of the United States. The limitations are industry-specific and are developed after consideration of the availability and economic achievability of the technology used as the basis for the limitations. The NPDES permitting requirements, and the requirements of Section 103 of the Marine Protection, Research, and Sanctuaries Act, also apply to disposal and discharge into the ocean.

EPA's issuance of new-source NPDES permits requires an environmental analysis under NEPA, and an Environmental Impact Statement, if required, to address the full range of potential impacts, including changes in land use, population density, or impacts on air quality, noise, wildlife, and other resources.

17.1.7 Safe Drinking Water Act

The Safe Drinking Water Act (SDWA) mandates the EPA to establish national drinking water standards for various contaminants. It also establishes the underground injection control (UIC) regulations that set minimum standards for siting, construction, operation, maintenance, and closure of injection wells. Injection wells are bored, driven, or dug holes used for the subsurface emplacement of fluids. Injection practices may be categorized as deep disposal of hazardous, industrial, or municipal wastewater; injection for enhanced recovery of oil and gas and disposal of brines and produced fluids; solution mining of minerals; or injection of hazardous or nuclear wastewaters above or into an underground source of drinking water. The purpose of the standards is to prevent injection wells from introducing contaminants into underground sources of drinking water that may cause a public water supply system to violate a National Drinking Water Standard or otherwise adversely affect human health and the environment. The permitting process addresses a range of environmental concerns.

17.1.8 Occupational Safety and Health Administration

The safety of workers is protected by laws and regulations governing public health in the workplace. The laws apply to normal operational activities and include all provisions for standard injury and illness prevention, construction requirements, and requirements for the handling of chemicals and prevention of infection and disease. Worker safety programs are industry-specific and ensure protection during normal operations and emergency conditions.

17.2 Hazardous Waste Studies

The analysis of potential hazardous waste impacts for a proposed project or action has three general steps requiring three separate reports: the initial site assessment (ISA), the preliminary site investigation (PSI), and remedial investigation/feasibility study (RI/FS).

17.2.1 Initial site assessment

The ISA study is a research activity on past and present land uses in the area or potential effects of the proposed project or action. There are three basic steps:

1. Regulatory agency record review
2. Site inspection and owner interview; review of historical records
3. Report preparation

The task begins with contact with the Environmental Protection Agency and appropriate state and local environmental and health regulatory agencies to identify any known hazardous waste sites in the area of potential project or action effect. Because of the various stringent legislative requirements discussed in the previous section, inventories of hazardous waste sites are comprehensive and up-to-date. EPA will provide the CERCLIS list and the National Priorities List. Requests for an "environmental audit research," government records report, or similar terminology from appropriate state or local agencies should yield a report covering the following:

- Permits to generate hazardous waste and/or handle hazardous materials and/or maintain an underground storage tank
- A log of reports of releases or threatened releases with potential for injury or groundwater contamination, by date
- A log of the agency hazardous materials team responses and incidents
- Underground storage tank removal permits
- Underground storage tank cleanup logs and record of tank leaks
- State list of identified hazardous waste sites
- State toxic substance control program information system
- EPA CERCLA site and event listing

Available information will include types of problems, the list of contaminants involved, status of investigations or cleanup, and recommended remediation. The state and/or local water resources and quality control boards or agencies should be contacted for groundwater quality information in test wells or nearby private or municipal wells.

Past and present land uses are determined through review of historic maps, coordination with local officials, site inspections, and interviews with property owners. For example, an ISA prepared for a project in San Bernardino, California, used Sanborn Fire Insurance maps from 1894, 1906, and 1950 and aerial photographs from 1965, 1972, 1979, and 1986 to determine past land uses in a project area for a proposed highway-widening project.

Field inspections to identify potential hazardous waste sites or materials will include a search for such elements as storage structures, pipelines, landfills, surface staining, oil sheen, odors, vegetation damage, or possible materials such as asbestos, paint, fireproofing, or pipe wrap. Existing land use and an interview with the owner or operator will yield additional information on possible storage structures, contamination, or hazardous materials.

The result of an ISA is a conclusion about the existence of a known or potential hazardous waste site. The report provides the basis for considering the next steps in the process. The best option is to review proposed alternatives to ascertain if the hazardous waste site can be avoided through changes in location, design, or characteristics. If not, the process proceeds to a preliminary site investigation, which can be time-consuming.

17.2.2 Preliminary site investigation

The purpose of the PSI is to determine whether hazardous waste is actually present through field testing, sampling, and laboratory analysis for contaminants. Testing normally includes soil samples, core borings, soil gas probes, and sometimes test wells to check for groundwater contamination. If hazardous waste is present, the investigation should identify the type and level of waste present, limit of contamination, and estimated costs for remediation.

Upon identification of a hazardous waste site, the appropriate state and local regulatory agencies should be notified with a request that the owner of the property be notified of the contamination. It is the responsibility of the property owner or other potentially responsible party to characterize the hazardous waste site and, where appropriate, to remediate the contamination.

The PSI report should include the detailed methodologies, studies, and remedial actions. The environmental document should contain a summary of the investigation, written in easily understood language. The environmental document should contain the following:

- A map showing the location of the sites
- A description of the contaminants, level of contamination, and federal and state maximum levels for the contaminants
- Alternatives that avoid the hazardous waste and why such avoidance is not practicable
- Potential remedial actions
- Results of coordination with appropriate regulatory agencies
- Status of investigation, including potential costs and time estimate to complete remedial actions
- The procedure to be followed should any hazardous waste or material not previously identified be discovered during construction

The above information for each proposed alternative becomes a consideration in the selection of alternatives. After an alternative is selected, the final environmental document should describe results of all coordination with regulatory agencies and whether the responsible party has been identified and has agreed to remediate the contamination.

17.2.3 Remedial investigation/ feasibility study

During the final design stages of the proposed project or action, efforts should be made to avoid or minimize impact with hazardous waste sites. A comprehensive investigation is required for all hazardous waste sites which cannot be avoided. The RI/FS determines the characteristics and extent of the contamination and develops a detailed hazardous waste management plan which recommends the most cost-effective method of remediation. Here are examples of remedial actions (Los Angeles County Metropolitan Transportation Authority 1993):

- *Containment barrier*—install vertical dike to block horizontal movement of contaminant.
- *Cap site*—install horizontal impermeable layer to reduce rainfall infiltration.
- *Excavate and dispose*—remove contaminated soil and dispose in an approved site.
- *Excavate and treat*—remove contaminated soil and treat (includes spreading and land farming).
- *Remove free product*—remove floating product from water table.
- *Pump and treat groundwater*—generally employed to remove dissolved contaminants.
- *Treatment at hookup*—install water treatment devices at each dwelling or other place of use.
- *Enhanced biodegradation*—use of any available technology to promote bacterial decomposition of contaminants.
- *Replace supply*—provide alternative water supply to affected parties.
- *Vapor extraction.*
- *Vent soil*—bore holes in soil to allow volatilization of contaminants.

A valuable source of information on remediation is the U.S. Geological Survey (USGS). The USGS provides technical assistance and research to other federal, state, and local agencies in many areas of expertise, as already discussed in Chap. 16.

USGS recent research in assessing the potential for contamination of groundwater and remediation of contaminated sites includes a modified groundwater flow model to determine vapor flow above the water table. The model can be used to determine optimal placement of wells and pumping rates to extract contaminants with the greatest efficiency. Another example is the USGS continual research on public health contaminants in groundwater, such as herbicides from central U.S. agricultural lands, and in surface water, such as the assistance in issuing a public health advisory in Washington state for fish consumption based on USGS studies of DDT in agricultural soil, stream water, stream sediment, and fish tissue in the Yakima River (U.S. Department of Interior, USGS 1995b).

The USGS also has conducted research on the use of natural and enhanced bioremediation techniques to degrade organic compounds soils. Microorganisms naturally found in soils are active consumers of fuel-derived toxic compounds and can transform the compounds into harmless carbon dioxide (U.S. Department of Interior, USGS 1995c).

If the remedial action recommended in the RI/FS is significantly different from the actions listed in the final environmental document, then a supplemental environmental document may be required.

17.3   Use of Scoping to Identify Issues and Concerns

Engineers, planners, and environmental health specialists will know of potential real risks likely to be associated with any particular proposed project or action. The abundance of strictly enforced regulations and permitting processes ensures against public health risks.

Just as important, however, is the public's perception of what the health and safety risks may be, whether the concerns are founded or unfounded in fact. It is therefore extremely important that the environmental document specifically address any issues and concerns raised by the public. Failure to recognize the importance of public perception can lead to catastrophic results in subsequent planning phases. The fact that a particular concern is raised by the public is, in itself, a reason for a thorough discussion within the environmental document. Applicable regulations and the incorporation of required prevention measures should be explained for each issue raised. In some cases, public input may reveal information on local conditions or historical events that would otherwise not be known by the environmental analysts.

As an example, the following environmental health and safety concerns were identified in the public scoping process for a proposed municipal solid waste landfill involving transport of the waste via rail

(U.S. Department of Interior, BLM, 1995):

- Hazardous materials in the waste stream
- Vectors transported with the wastes
- Fires at the landfill
- Use of leached-ore residue for landfill cover
- Interference with aircraft using a nearby aerial gunnery range
- Contamination of groundwater
- Transport train delays
- At-grade railroad crossings
- Train accidents
- Highway safety on local roads during peak tourist season

Each of these issues was addressed in the environmental studies and discussed within the Draft Environmental Impact Statement.

17.4 Mitigation

Discussions of environmental health and safety issues within an environmental document normally consist of a description of mitigation measures. The numerous regulatory and project design measures incorporated into the project will most likely lead to a conclusion of no significant impact with installation and implementation of mitigation measures. Individual measures are far too numerous, complex, and project-specific to be listed in this text, but descriptions should be included within the environmental document for a particular project in sufficient detail to offset public concerns.

Chapter
18

Water Resources

In public opinion polls, water has been noted by many residents of the United States as the most important environmentally sensitive natural resource. The nation's watersheds, streams, rivers, groundwater supplies, and marine environments are protected by numerous legislative and regulatory provisions. This chapter presents some of the most important issues in assessing potential environmental effects of a proposed project or action. The analyst's best tool in the effective impact assessment is early and productive coordination with appropriate federal, state, regional, and local agencies. Early coordination and consultation facilitate identification of project-specific, relevant information and issues and enable the analyst to focus studies on those areas warranted by the potential magnitude of the water resources impacts.

18.1 Groundwater

The study of groundwater resource impacts is closely related to studies of geology and soils. Productive interaction of various members of the multidisciplinary team is necessary to ensure that all important characteristics are noted and all possible impacts are considered.

18.1.1 Existing characteristics

Relevant information on groundwater characteristics of a study area can be determined through coordination with local water management districts, water quality agencies, or municipal water suppliers. State Stormwater Management Acts require watershed management plans or water quality control plans for groundwater basins. These plans will identify particularly sensitive groundwater issues. Review of local or regional water quality control or management plans will indicate most recent data on water supply, volume, flow, quality, use, and future goals and objectives. Extensive information on surface and groundwater flow and quality also is available from the USGS.

The USGS provides water supply and quality information through a national assessment program designed to study water quality in representative basins covering more than 50 percent of the United States (U.S. Department of Interior, USGS 1995a). The USGS-developed mathematical model of groundwater flow has become a widely used computer-based model in the groundwater industry (U.S. Department of Interior, USGS 1995e). For example, USGS programs in California (U.S. Department of Interior, USGS 1995i)

- Assess water resources quantity and quality throughout the state

- Assist water management agencies in southern California in the study of using aquifers for storage and use of reclaimed wastewater for groundwater recharge and irrigation

- Develop water management programs to characterize hydrogeologic conditions in areas that appear to have good potential for water banking (recharging groundwater systems for future pumping), and to optimize combined use of surface and groundwater, control of water levels in an urban area subject to liquefaction during a major earthquake, and containment of groundwater pollutants from Environmental Protection Agency Superfund sites

- Develop management strategies for controlling seawater intrusion into coastal aquifers

The USGS data are stored in the national Water Data Storage and Retrieval System (WATSTORE), which includes a daily-values file that contains 300 million observations of stream flow, water quality, sediment discharge, and groundwater-level data; a water quality file that contains 4.1 million surface and groundwater analyses; a peak-flow file that contains nearly 600,000 observations of annual peaks of stream flow and river stage; and a groundwater site inventory file that contains information for more than 1.4 million wells. Much of this information has been made available over the Internet (U.S. Department of Interior, USGS 1995d).

The USGS also operates the National Water-Use Information Program used to collect, store, analyze, and disseminate water-use information nationally and locally to a wide variety of government agencies and private organizations (U.S. Department of Interior, USGS 1995g).

Water quality control plans identify uses of groundwater supplies, such as municipal and domestic, agricultural, industrial services, and/or industrial process. Use characteristics within the basin are described. Specific water quality objectives, or standards, are compared with recent and historical quality testing results from monitoring wells.

The major water-bearing geologic formations (aquifers) are identified as related to soils and geologic information. Flow and quantity characteristics of groundwater resources are dependent on local surface and geologic conditions. In areas of faults and historic fault activity, impermeable vertical and horizontal restrictions to groundwater flow may occur. In limestone areas, subsurface caverns and streams may determine groundwater characteristics.

Natural and/or created groundwater recharge areas should be noted on mapping. A *recharge area* is the area in which water reaches the zone of saturation (groundwater) by surface infiltration. Information will likely be available in local or regional water management plans on the principal sources of groundwater recharge. If the source is precipitation, how much of the average annual rainfall is available for recharge? Other sources may include subsurface inflow, seepage from streams and rivers, or artificial recharge. In some areas of California, for example, recharge facilities known as *conservation basins* have been constructed. These basins are filled through use of diverted local stream flow and from water imported through the California Aqueduct State Water Project. Drywells also can be used to percolate drainage water.

Other information on groundwater recharge may be available from county or regional flood control or water conservation districts. A local or regional hydrology manual will contain specific hydrologic soil classifications based on infiltration rates.

18.1.2 Groundwater impacts and mitigation

Effects on groundwater may be direct, indirect, physical, or chemical. Direct, physical impacts could include loss of wells, interception of the water table, or other physical changes through earthwork, blasting, etc. that alter flow, recharge, or other hydrologic conditions. Depending on the type of project proposed, a physical interception of the water table may, in turn, require continual dewatering through pumping. Dewatering can cause localized drawdowns of the water table elevation and adversely affect local wells.

Substantial new demand and use of groundwater can cause local or regional drawdowns. For example, a new housing project with individual residential wells, or a system of community wells, may adversely lower the water level and available groundwater supply at nearby existing housing developments that also depend on residential wells. If the water level is lowered by deeper new wells below the level of existing wells, those residences on existing wells will be without water. Figure 18.1 illustrates how groundwater wells can affect

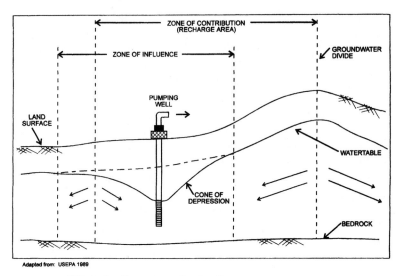

Adapted from: USEPA 1989

Figure 18.1   Effect of wells on groundwater flow.

the flow of groundwater by lowering water levels in an area around the well, known as the *zone of influence* or *cone of depression*. The full recharge area of a well is often called the *zone of contribution*.

Of special concern in coastal areas is the potential for urban demand on groundwater to lower the existing water table to a degree to upset the subsurface balance in flow between freshwater and salt water. An adverse infiltration of salt water into underground freshwater supplies can produce serious regional impacts.

Another potential impact of groundwater withdrawal in particular geographic areas is increased risk of land subsidence. Land subsidence due to pumping of groundwater occurs in nearly every state (U.S. Department of Interior, USGS 1995e).

Another factor affecting groundwater supplies is the creation of impervious surfaces within the recharge area. This type of impact occurs with projects involving large areas of pavement or a significant number of buildings. By covering the recharge area with impermeable surfaces, the projects reduce the total area available for water percolation through the soils to underlying aquifers. The same impact would occur for flood control channels that are entirely paved. For this reason, many artificial river and stream channels are constructed with paved sides and natural bottoms.

A common approach to evaluation of potential recharge-area impacts is to compare the created impervious surface with the total for the watershed. Unfortunately, the same argument is sometimes made for two entirely different situations. One logic used is that the existing watershed is so large and undeveloped, with such a large pervious recharge area, that a small increase in impervious surface would not significantly affect the quantity of groundwater percolation. The other logic is that the watershed is already urbanized to such an extent that the increase in impervious surface, when compared with the total surfaced area, would add an insignificant contribution. So it seems that the two extremes of developed and nondeveloped watersheds would not be affected by small amounts of new impervious surface, and that the only impacts would occur in partially developed watersheds.

Although the logic above is most likely applicable for the two cases stated, the analyst must also consider the possible cumulative effects of a particular project or action. Groundwater recharge areas are seldom lost all at once. Small amounts of impervious surface are added little by little, each with no significant impact, until there is a significant overall effect. Whenever possible, even the smallest project should attempt to incorporate detention ponds or other mitigation measures to detain water and permit maximum recharge within the remaining pervious surface.

An increase in impervious surface can contribute to groundwater quality impacts as well as quantity. Possible water quality effects on groundwater may include increases in suspended solids from erosion or in chemical contaminants through rapid runoff from impervious surfaces. If runoff picks up pollutants such as herbicides, fertilizer, oil, gas, metals, organic or inorganic compounds, or deicing salts, these can move more rapidly into groundwater aquifers in higher concentrations.

The possibility of groundwater contamination can be increased by projects that require excavation within sensitive geologic areas. Geologic characteristics of a particular area, such as rock fractures, sinkholes, solution channels, or shallow soils, may contribute to a direct flow of contaminants into local aquifers. Such an impact would be most critical in areas of low groundwater recharge and yield, where dilution and flushing would be minimal.

Mitigation of potential groundwater impacts is best approached through coordination with appropriate officials and an analysis of consistency with water conservation plans and water quality control plans. Geologic voids can be filled or capped, and soil cover can be added to areas of exposed rock. Other measures would apply to both groundwater and surface water quality and may include such project-specific techniques as installation of oil and grease separators in large parking lots. Vegetated swales, drainageways, and created wetlands with extended detention also may be used to filter sediments and contaminants from drainage.

As with surface waters, groundwater resources may be put at increased risk by certain projects to catastrophic events, such as spills of contaminants or hazardous materials. Each proposed project or action should be examined in terms of the probability of such an occurrence versus the magnitude of the possible impact. Mitigation may be to require maintaining cleanup equipment on site or to establish special emergency response teams.

When fuels, pesticides, fertilizers, sewage, solvents, and other substances enter the soil, the contaminants percolate through the soil to groundwater resources. Mitigation may be in the form of using computer models to monitor groundwater flow to determine the optimal placement of wells and pumping rates to extract contaminants.

Another form of mitigation, particularly for sites that may have previously been contaminated, is bioremediation. The Superfund legislation (CERCLA) prompted the authorization of the Toxic Substances Hydrology Program by the USGS. The program consisted of systematically investigating the most important categories of wastes at sites throughout the United States. One of the principal findings was that microorganisms in shallow aquifers affect the fate and transport of virtually all kinds of toxic substances (U.S. Department of Interior, USGS 1995b). Microorganisms naturally present in the soils can actively consume fuel-derived toxic compounds and transform them to harmless carbon dioxide. The rate of biodegradation can be greatly increased by stimulating the natural microbial community through the addition of nutrients.

Examples of successful bioremediation projects (U.S. Department of Interior, USGS 1995b) are summarized below to show the types of pollution that can occur and the remediation techniques.

A pipeline carrying crude oil burst and contaminated the underlying aquifer. The natural microbial population caused the plume of contaminated groundwater to stop expanding after a few years, without human intervention.

A sewage effluent plume due to disposal of sewage effluent in septic drain fields caused nitrate contamination in a shallow aquifer. Denitrification was rapidly accomplished by microbial populations.

Chlorinated solvents have been successfully used by microorganisms as oxidants, which remedies solvent contamination of groundwater.

Biological and nonbiological processes have been shown to degrade pesticide and nitrogen fertilizer contamination.

Studies of gasoline-contaminated sites have shown the importance of processes in the unsaturated zone (the zone above the water table) in degrading contaminants.

In Florida, creosote and chlorinated phenols leaked to the underlying aquifer through unlined ponds and were transported toward

Pensacola Bay. Studies showed that microorganisms can adapt to extremely harsh chemical conditions and that microbial degradation was restricting migration of the contaminant plume.

18.1.3   Sole-source aquifers and other special conditions

Sensitive aquifers and the specific recharge areas and characteristics should be identified. A sole-source aquifer is an aquifer which is the sole or principal drinking water source for an area and which, if contaminated, would create a significant hazard to public health. Sole-source aquifers are designated and protected under Section 1424(e) of the Safe Drinking Water Act. Any designated critical aquifer protection areas (CAPAs) within the sole-source aquifer or recharge area in the area of potential effect of the proposed project or action should be identified. Early coordination regarding sole-source aquifers is required with the Environmental Protection Agency, which will review proposed alternatives and offer opinions on possible impacts. A separate report is normally prepared at the draft environmental document stage when other technical analyses are being conducted. The potential impacts of all proposed alternatives on the sole-source aquifer should be comparatively discussed. If the selected alternative involves a sole-source aquifer, the final environmental document must contain appropriate analysis and consultation results to ensure that the proposed project or action will not contaminate the aquifer (40 CFR Part 149).

Other regulatory requirements that may apply include wellhead protection areas as authorized by the 1986 amendments to the Safe Drinking Water Act. Each state must develop wellhead protection plans to protect groundwater that supplies wells and well fields that contribute drinking water to public water supply systems. The selected alternative of the proposed project or action must comply with these plans. Such documentation must be included in the final environmental document.

In some areas where agricultural land use is dominant, pest management zones may have been established. These designated zones are areas where pesticides have been found in groundwater and thus future spraying is prohibited.

18.2   Municipal Water Supply and Wastewater Systems

In urban and suburban areas, it is important to determine the major source of potable water: a municipal system or local wells. The source of municipal supply systems and whether the water quality from active wells or surface waters meets standards, or is treated to do so, should be noted. Public water supply standards are established by the Environmental Protection Agency as authorized by the Safe Drinking Water Act. Contained in 40 CFR Parts 141 to 143, the National Primary and Secondary Drinking Water Regulations establish nonenforceable health goals [maximum contaminant level goals (MCLGs)] and enforceable maximum contaminant levels (MCLs) for numerous organic and inorganic pollutants. In rural areas, the primary aquifer source for wells should be described in terms of yields and physical and chemical characteristics.

The wellhead protection program of the state will identify wellhead protection areas, sources of contaminants, management approaches, contingency plans, and other valuable information on groundwater supplying public water supply systems.

Area wastewater systems also should be identified, including regional facilities' plans, particular characteristics of treatment plants serving the proposed project or action, and current capacity compared with quantities. Wastewater reuse within the basin may be important depending on the type of project being assessed. As discussed in Chap. 17, wastewater treatment plants are required as part of the Clean Water Act NPDES permit to meet established effluent quality standards. Requirements consist of both numerical values and periodically reporting the average or maximum values.

18.3 Surface Water

Many of the potential surface water impacts (Fig. 18.2) will be the same as potential groundwater impacts, such as a possible increase in suspended solids from erosion or chemical contamination from runoff. The same is true for subsequent discussions in this text related to floodplains, coastal areas, and wetlands. Surface water studies also include consideration of the aquatic biotic community and of recreational and commercial uses.

18.3.1 Existing characteristics

In addition to the sources of information discussed in the section on groundwater, surface water quality and aquatic community data may be available from the U.S. Department of Interior, Fish and Wildlife Service, state departments of environmental resources; or fish and game, river basin commissions, or local boating and fishing organizations. The USGS information on surface waters may include results of chemical and biological testing as well as detailed flow information collected at gaging stations. For example, surface water discharge (flow) data was collected by the USGS at 10,240 stations in 1994; surface

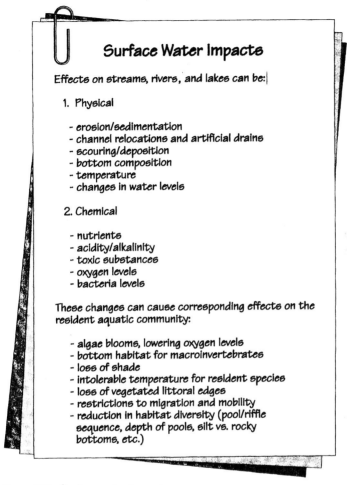

Figure 18.2 Surface water impacts.

water quality data was collected at 3098 stations (U.S. Department of Interior, USGS 1995d).

Water quality standards for surface waters are normally set by state or local agencies and are based on use classifications. Examples of uses include water contact recreation and aquatic life; natural trout waters; and recreation (stocked) trout waters. Depending on the designated use for a particular section of stream or river, maximum permitted concentrations of water quality pollutants will be established, such as acidity (as measured in pH), alkalinity, biological oxygen demand (BOD), chemical oxygen demand (COD), dissolved oxygen (DO), dissolved solids (DS), nutrients (nitrates and phosphates), fecal coliforms, and toxic substances (heavy metals, etc.).

As appropriate based on the results of scoping, the environmental document should include a mapping of major surface waterways, water bodies, and watersheds. Existing data should be summarized and compared with applicable water quality standards. Historic trends and sources of pollutants should be described. For proposed projects or actions requiring detailed water resource studies, information is usually contained in a supporting technical report and summarized in the environmental document.

For proposed projects or actions where surface water impact is a key issue and no available data exists for local resources, a stream or lake survey may be conducted by the study team. Such surveys include sampling over a period of time (one year is the best minimum to ensure all seasons) of water quality and aquatic life. Physical conditions noted at each sampling station include such factors as bottom type, bank cover, percentage of shade, pool and riffle sequence, water temperature, and pH. Water samples are collected for analysis in the laboratory for water standard parameters. Bottom-dwelling (benthic) macroinvertebrates also may be systematically collected. Benthic macroinvertebrates, such as tube worms, mollusks, and larvae of stoneflies, mayflies, caddisflies, and beetles, are commonly used indicators of water quality. The established benthic community—species composition, quantity, and diversity—yields an abundant amount of information on water quality and condition of streams, rivers, and lakes. Fish sampling also may be included through use of seining or electroshocking.

The presence of any state or federal endangered, threatened, or designated sensitive aquatic plant or animal species should be identified through coordination with the Fish and Wildlife Service, National Marine Fisheries Service, and appropriate state agencies.

18.3.2 Surface water impacts and mitigation

The ability to assess potential effects on surface water resources requires a thorough understanding of the physical and operational characteristics of the proposed alternatives of the project or action. The analyst must be knowledgeable about the construction phase, operational phase, and exactly what physical changes to existing surface water resources are a component of the alternatives. Impacts can then be comparatively assessed based on the level of design detail available.

Construction (short-term) effects. The primary impact associated with construction activities is usually erosion and sedimentation effects. Stream and lake sedimentation and turbidity loading can cover bottoms with silt and destroy benthic organisms. This in turn can eliminate food supplies for particular species and totally upset the balance of the aquatic ecological system.

Some projects may require direct in-channel work during construction activities, or alteration of the surface water channel or shoreline. Channel relocations are more specifically addressed in the next section of this chapter. Direct channel disturbance results in a temporary loss of habitat and bottom-dwelling species. Compaction of stream bottom habitat by construction equipment can permanently degrade the ability to support aquatic vegetation or bottom-dwelling organisms.

The other impact associated with construction activities is that runoff from construction sites will contain water quality contaminants. Runoff impacts are further discussed in the long-term impacts section of this chapter. The probability of spills of hazardous materials directly into water bodies should also be considered.

As discussed in greater detail in Chap. 16, installation of proper erosion control measures can normally mitigate construction-related erosion impacts successfully. Often a requirement is that final design of the selected alternative contain specific erosion and sedimentation control plans. These plans undergo a series of approvals and permits depending on the specific state or local requirements.

Another example of required procedures to minimize impact is a water construction permit that may limit the time of year for any in-stream work, based on the feeding, nesting, or breeding requirements of sensitive aquatic species. Equipment stream crossings should be conducted with minimal disturbance, on rock fills with many pipes to permit natural water flow. Removal should return the bottom to natural compaction characteristics.

The Clean Water Act requires preparation and submission of a general construction activity stormwater permit before construction is begun. The permit requires preparation of a stormwater pollution prevention plan. The plan is based on the use of best management practices (BMPs). BMPs applicable to construction sites include measures to prevent erosion, prevent pollutants from the construction material from mixing with stormwater, and trap pollutants before they can be discharged.

The plan also contains requirements for the construction contractor to prepare and implement a hazardous materials management plan to reduce the possibility of chemical spills or releases to drainage channels. Proper material handling, storage, and disposal protocols are established and enforced.

Channel relocations. Long-term changes to course, current, or the cross section of the channel or its floodplain may be included as part of the proposed project or action. Sometimes exact design characteristics may not be known prior to the final design stages. The comparative analysis of proposed alternatives at the draft environmental document stage should be accomplished with as much detail as possible, however, to permit inclusion of this important potential impact in the decision-making process.

Many miles of streams in the United States have been lost due to stream channelization. Old practices often consisted of straightening streams into short sections of totally paved channels or culverts. The environmental impact assessment process over the past decades and the agency coordination it has fostered have made enormous changes in the manner in which stream relocations are conducted.

Stream channelization often is associated with linear projects, such as highways or railroads, with an engineering need to occupy relative flatlands. In hilly or mountainous areas, the only relative flatland may be that associated with a stream and its floodplain. Often the comparison of effects of valley floor alternatives with hillside alternatives centers on the relative damage of large, destructive cuts and fills on the hillsides versus fill and stream relocation, floodplains, or wetland impacts on the valley bottom.

The assessment of stream channelization impacts normally begins by comparing the length of the natural stream to be destroyed with the length of the new channel. The stream channel to be destroyed is surveyed, noting important physical, chemical, and biological features. The most obvious impact is the loss of all aquatic habitat, streamside vegetation, and benthic organisms within the destroyed channel. This impact should be quantified within the environmental document as a direct impact of the proposed project or action. The loss should be evaluated based on the sensitivity of the stream, aquatic habitat, endemic species, and designated uses.

In all cases, the goal should be to design the new channel to maintain the same stream length. A loss of length and meanders into a relatively straight channel creates disturbances to the natural dynamics of water flow. Velocity will increase, causing upstream scouring and downstream deposition. Channel sides will be under constant erosion pressure as the stream flow tries to reach its former equilibrium. Changes in velocity can be estimated based on comparison of old and new channel lengths and flow characteristics of the stream. Degree of upset to the natural stream flow dynamics can then be estimated.

Discharge of dredge or fill material within streams, wetlands, or other waters of the United States requires a U.S. Army Corps of Engineers (Corps) Section 404 (Clean Water Act) permit. Details of the permit application information and process are contained within 33 CFR Parts 320 through 330. Section 404 permits are further discussed in Sec. 18.4 of this chapter. Other permits or regulations that may apply are those of the Rivers and Harbors Act of 1899. Under this act, Section 9 permits apply to crossings of navigable waters of the United States. Section 9 permits are issued by the Corps for dams and dikes and by the Department of Transportation, U.S. Coast Guard, for bridges and causeways. Section 10 of the act requires a permit for various types of work performed in navigable waters, including stream channelization, excavation, and filling.

Sections 9 and 10 of the Rivers and Harbors Act apply to navigable waters of the United States, while Section 404 of the Clean Water Act applies to waters of the United States. Waters of the United States mean more than navigable waters—they include floodplains and wetlands.

The Fish and Wildlife Coordination Act requires consultation with the Fish and Wildlife Service and the state agency responsible for wildlife resources for any modification to a stream channel or other body of water. The environmental document should contain a description of the uses of the stream or body of water and an analysis of impacts to fish and wildlife resulting from loss, degradation, or modification of habitat.

Mitigation of stream channelization impacts includes such measures as ensuring no net loss in stream length (that is, new channel same length as natural channel); natural bottoms as opposed to concrete channels; shape of concrete channels (trapezoidal) to maintain low-flow conditions; wide channel bottoms to permit stream to cut its own meandering stream course; and stabilization of outside curve banks with rock placements.

Rocks and gravel can be placed randomly within the new channels to encourage rapid naturalization of the streambed and development of a pool and riffle sequence. Stream banks should be stabilized before diverting the flow of the stream from the old to the new channels.

Continuing (operational) impacts.   Erosion and sedimentation effects should not be ignored as potential continuing impacts of a proposed project or action, particularly if the action is a management plan for large geographic areas or a land-use plan for municipal areas. The evaluation of alternatives for long-term plan components must include an assessment of the possible secondary effects of long-term permitted uses, such as timbering techniques, grazing, mining, or continued growth in housing and development. Such activities can cause loss of vegetation and exposure of bare ground within watersheds and associated erosion and stream sedimentation effects.

Potential long-term erosion and sedimentation effects can best be mitigated through incorporation of effective control and restoration requirements within the planning documents. These required mitigation measures should be applicable to all permitted activities through the use of permits or approvals prior to project implementation. Mitigation, to be effective, must include proof of funding sources for implementation of the committed measures.

Continuing, long-term water quality effects are mostly related to discharges of runoff from paved surfaces. An increase in impervious strata produces a proportionate increase in the amount of runoff carrying pollutants. This type of impact is discussed in Sec. 18.1.

Highway projects are used as an example of types of impact assessment required. Stormwater runoff containing vehicle-generated highway pollutants enters local drainage systems and, at bridges, can directly enter streams, rivers, or lakes.

Several studies have been conducted to develop methodologies for predicting highway runoff impacts on surface water quality (U.S. Department of Transportation, FHWA 1985; Lord 1987; California Department of Transportation 1982). For example, highway runoff constituents shown to be statistically correlated to traffic have been identified as particulates, total nitrogen, lead, zinc, and chemical oxygen demand.

Methodologies include use of formulas and assumptions on pollutant deposition per vehicle, rainfall amounts, and runoff coefficients for the paved surface. A worst-case analysis assumes that highway runoff is discharged directly into a receiving stream, with no intervening soil, vegetation, or dilution from convergence with additional sources. Calculations can be made by entering data on average daily traffic volumes and area of paved surface. Results, after several stages of calculation, will be in flow-weighted concentrations in milligrams per liter for various pollutants, such as lead, zinc, filterable solids, chemical oxygen demand, and total nitrogen. These concentrations can be directly compared to effluent standards. The results could also be refined based on the volume of water in the receiving stream, the dilution factor, and compared with water quality standards established for the stream or river for protection of human and aquatic life, and based on designated uses.

Water quality limits set in the NPDES permit provide the basis for application and enforcement of surface water quality standards and objectives. The NPDES permit requires monitoring (testing) of the effluent and of the receiving waters upstream and downstream of the discharge.

Accidental spills of hazardous materials are always a risk with transportation systems, including highways, railroads, and navigation systems. As discussed in the section on groundwater impacts, the possible magnitude of impact of such spills can be devastating, as demonstrated by coastal and marine life impacts of oil spills from tankers. The probability of such spills, however, is low.

Water quality impacts are not necessarily associated with paved surface runoff. Runoff from agricultural lands or such uses as golf courses can transport fertilizers and pesticides to receiving streams and rivers. Grazing or dairy land uses can contribute significant amounts of nutrients and animal coliform bacteria.

Industrial uses can contribute toxins or heavy metals directly into surface waterways. Some processing plants or nuclear reactors can emit effluent with significantly higher temperatures than the receiving waters, causing corresponding impacts on aquatic life.

Another important potential impact on surface waters is the possibility of secondary and/or cumulative impacts. If the proposed project or action places stress on a particular geographic area for growth, water quality impacts related to that future growth must be assessed to the degree possible with available information. If a project will facilitate and accelerate the rate of development and suburbanization, such growth in turn will increase impervious surface, accelerate discharges of polluted runoff, and increase stream channel erosion and sedimentation. Pressures may also be placed on existing treatment plants or water demand facilities.

Water quality impacts of runoff can be mitigated through use of stormwater runoff regulations. Stormwater management practices under such regulations may include

- On-site infiltration
- Flow attenuation by open vegetated swales and natural depressions
- Stormwater retention structures
- Stormwater detention structures

These measures can significantly reduce pollutant loads and control runoff.

Mitigation measures for stream or watershed impacts may include restoration projects for nearby, previously degraded watersheds or streams.

A Clean Water Act Section 401 water quality certification, issued by the appropriate state agency, is required for proposed projects or actions potentially affecting water quality. The certification ensures compliance with established effluent guidelines and standards.

Recreational and commercial uses. The evaluation of potential surface water impacts must include consideration of recreational and commercial uses and the possible impact of the proposed project or action on such uses. Existing uses should be described and quantified to the extent possible. Impact assessment should include such items as boating and passive access, hunting restrictions (waterfowl), aesthetic qualities, and possible degradation of fisheries or other commercial products through water quality pollution. Coordination with appropriate state and local agencies, sporting clubs, etc., should be documented.

18.3.3 Wild and scenic rivers

The Wild and Scenic Rivers Act provides for preservation of free-flow conditions of certain selected rivers of the nation and protection of those rivers' immediate environments for the benefit and enjoyment of present and future generations. To be eligible for the National Wild and Scenic River System, a river must be free-flowing and must possess outstandingly remarkable scenic, recreational, geologic, fish and wildlife, historic, cultural, or other similar values. The act required river studies to be conducted by the Department of Interior (Fish and Wildlife Service, Bureau of Land Management, or National Park Service) or Department of Agriculture (Forest Service) for all rivers, or river segments, potentially eligible for the system. Eligible river segments are classified according to the extent of evidence of human activity as wild, scenic, or recreational. Specific eligibility criteria apply to each classification. Each designated section of river on the system has a management plan, including principles, kinds and amounts of public use, and specific management measures.

The act protects rivers listed on the Nationwide Rivers Inventory and those that may be potentially eligible for listing. Coordination is required prior to any action that may foreclose a river's inclusion in the system.

Follow these steps to comply with the Wild and Scenic Rivers Act:

1. Determine whether the proposed project or action could affect an inventory river by checking the current regional inventory lists.

2. Assess potential adverse effects on the natural, cultural, and recreational values of the inventory river segment. *Any action which could alter the river segment's ability to meet the eligibility and classification criteria should be considered an adverse impact. Actions which diminish free-flowing characteristics or actions which increase the degree of evidence of human activity, that is, level of development, could prevent qualification or change classification. Adverse effects may occur under conditions that include*

   - *Destruction or alteration of all or part of the free-flowing nature of the river*
   - *Introduction of visual, audible, or other sensory intrusions which are out of character with the river or alter its setting*
   - *Deterioration of water quality*
   - *Transfer or sale of property adjacent to an inventoried river*

3. Determine whether the proposed project or action could foreclose options to classify any portion of the segment as wild, scenic, or recreational river areas. *This mostly refers to actions that may cause a downgrade in classification based on increased evidence of human activity.*

4. Incorporate avoidance and mitigation measures into the proposed project or action to the maximum extent feasible.

The draft environmental document must include evidence of coordination with the agency responsible for managing the listed or study river (Fish and Wildlife Service, Bureau of Land Management, National Park Service, or Forest Service). If an adverse impact is anticipated, the draft environmental document must fully consider avoidance and mitigation measures for each alternative that may affect a segment of designated or proposed scenic river. If a selected alternative has potential adverse effects, the Finding of No Significant Impact or Final Environmental Impact Statement must identify committed measures to avoid or mitigate adverse effects, with documentation of concurrence with the managing agency.

Publicly owned segments of designated wild and scenic rivers are also protected by Section 4(f) of the DOT act of 1966. This act applies to transportation use of park, historic, and recreational land and is discussed in Chap. 10.

18.4   Section 404 Permits

Section 404 of the Clean Water Act establishes a program to regulate the discharge of dredged and fill material into waters of the United States, including wetlands. The Section 404 regulatory permit program is administered jointly by the EPA and the U.S. Army Corps of Engineers (Corps). The Corps issues the actual permit. Whereas the Section 404 permit applies to waters of the United States, the Corps also issues Section 9 and Section 10 permits under the Rivers and Harbors Act for activities applying to navigable waters of the United States.

The Corps Section 404 permit and the Section 9 and Section 10 permits review includes significant consideration of environmental impacts; among those are conservation, economics, aesthetics, wetlands, historic properties, fish and wildlife values, flood hazards, floodplain values, land use, navigation, shore erosion and accretion, recreation, water supply and conservation, water quality, energy needs, safety, food and fiber production, mineral needs, and considerations of property ownership.

The Environmental Protection Agency has issued Section 404(b)(1) guidelines that prohibit discharge of dredged or fill material

1. If there is a practicable alternative with less adverse impact on the aquatic environment unless the alternative poses other significant environmental problems

2. If the discharge will have an unacceptable adverse impact, whether individually or cumulatively, on the aquatic ecosystem

3. If the discharge will violate state water quality standards, violate toxic effluent standards, jeopardize a species listed as threatened or endangered under the Endangered Species Act, or violate any requirement of a marine sanctuary designated under the Marine Protection, Research, and Sanctuaries Act

4. Unless appropriate and practicable steps have been taken which will minimize potential adverse impacts of the discharge on the aquatic ecosystem

These are the four basic restrictions to permitting discharge of fill or dredged material. The guidelines require identification of all direct, indirect, secondary, and cumulative impacts which could result from a proposed discharge. All practicable steps must be taken to minimize the adverse impacts, including providing compensation (for example, wetland restoration and creation) for unavoidable impacts.

The Section 404(b)(1) guidelines (40 CFR Part 230) offer extensive and valuable information and methodologies to the environmental analyst for evaluation of potential environmental effects on water resources and aquatic ecosystems. The guidelines require factual determinations, in writing, of the potential short-term or long-term effects of a proposed discharge of dredged or fill material on the physical, chemical, and biological components of the aquatic environment. Determinations must be made regarding

- Physical substrate
- Water circulation, fluctuation, and salinity
- Suspended particulates and turbidity
- Contaminants
- Aquatic ecosystems and organisms
- Disposal sites
- Cumulative effects on the aquatic ecosystem
- Secondary effects on the aquatic ecosystem

Specific guidelines are given to assist in making the required factual determinations. Figure 18.3 lists the areas of possible impact that are discussed. The next step in adherence to the Section 404(b)(1) guidelines is to make findings of compliance or noncompliance with

1. Potential impacts on physical and chemical characteristics of the aquatic ecosystem
   - Substrate
   - Suspended particulates and turbidity
   - Water
   - Current patterns and water circulation
   - Normal water fluctuations
   - Salinity gradients

2. Potential impacts on biological characteristics of the aquatic ecosystem
   - Threatened and endangered species
   - Fish, crustaceans, mollusks, and other aquatic organisms in the food web
   - Other wildlife

3. Potential impacts on special aquatic sites
   - Sanctuaries and refuges
   - Wetlands
   - Mudflats
   - Vegetated shallows
   - Coral reefs
   - Riffle and pool complexes

Figure 18.3 Section 404(b)(1) guidance on evaluation of impacts for factual determinations.

the four restrictions noted above. Three findings are possible: (1) complying, (2) complying with inclusion of appropriate and practicable discharge conditions to minimize pollution or adverse effects on the affected aquatic ecosystems, or (3) not complying.

Section 404 permits are issued in two basic forms: individual permits and general permits. Individual permits are project-specific and are issued after a case-by-case review of individual permit applications. General permits authorize categories of activities in specific geographic regions or nationwide. If an activity is covered by a general permit, an application for a permit is not required. The general permits are issued when (1) proposed activities are substantially similar in nature and cause only minimal individual and cumulative environmental impacts; or (2) the general permit would result in avoiding unnecessary duplication of regulatory authority exercised by another agency. Based on the above, there are three types of general permits: regional, nationwide, and programmatic. An activity is authorized under general permits only if that activity and the permittee satisfy all the general permit's terms and conditions. Details of the Nationwide Permit Program are contained in 33 CFR Part 330.

The Section 404 process has a 30- to 45-day required public review and comment period. It is advantageous to integrate the Section 404 review process with the NEPA review process of the Environmental Assessment or Draft Environmental Impact Statement. Specific required details for actual permit application will most likely not be available at the draft environmental document stage, because numerous alternatives are being considered. The draft document should, however, note the location of proposed dredge and fill activities, the potential adverse effects, and proposed mitigation measures for each alternative. Documentation of coordination with the Corps and appropriate federal, state, and local resource agencies should be included. The final environmental document should identify, for the selected alternative, the location of permitting activities, quantities of dredge and fill material, potential impacts, and mitigation measures. Any outstanding unresolved issues should be identified.

The actual application for a Section 404 permit will likely be made after the final environmental document because of the need for detailed final design information. Some agencies may choose, however, to complete the permit at an earlier stage. In many cases, Section 404 permits may apply to projects or actions that are not "major federal actions" and therefore would not require an Environmental Impact Statement.

A Corps decision on a permit application requires either an Environmental Assessment or an Environmental Impact Statement, unless it is included within a categorical exclusion. If another federal agency is the lead agency for an action, the Corps will most often become a cooperating agency for the environmental document.

18.5   Marine Environment

Actions or projects that may affect marine resources will require an analysis of potential physical, chemical, and biological impacts. Such projects may include sewage treatment plants or industries with discharges into the ocean; offshore drilling projects; harbor developments or improvements; changes in shipping operations or type and quantity of activities at an existing port; or construction of bridge or tunnel crossings of bays or estuaries.

The discharge of effluent into ocean waters requires a Clean Water Act National Pollutant Discharge Elimination System (NPDES) permit. The permit establishes discharge limitations for all major wastewater constituents, aquatic life toxicants, noncarcinogens, and carcinogens. The NPDES permit also requires an effluent monitoring program and a receiving water monitoring program. Monitoring programs that may be in place because of NPDES permit requirements are a valuable source of information on the existing physical, chemical, and biological characteristics of the marine environment in an area of a proposed project or action.

Other important sources of information are state, regional, and local agencies and established ocean plans. The U.S. Geological Survey conducts extensive studies and monitoring of many of the nation's estuaries, bays, deltas, and marine environments, including water quality, water quantity, and such effects as saltwater intrusion into groundwater supplies. The Department of Commerce, National Marine Fisheries Service, will supply information on threatened and endangered marine species of plants and animals.

The description of the existing marine environment should be given as necessary to determine impacts and resolve issues relevant to the specific project or action being proposed and evaluated. The analyst is warned not to include extraneous data and thereby create an encyclopedic narrative just because the information may be interesting to the researcher. If the information and data are not relevant to the magnitude of potential effects and to the decision at hand, leave them out.

For the purpose of showing the types of information that may be relevant, the subjects addressed in an Environmental Impact Statement for a major municipal wastewater treatment facilities plan, with discharges into the Pacific Ocean off the coast of California (County Sanitation Districts of Los Angeles County 1994), are shown in Fig. 18.4.

Because of the direct and potentially significant impact on marine resources, the assessment outlined in Fig. 18.4 included a comprehensive description of existing chemical conditions, aquatic resources, physical characteristics, and historic pollution problems. This degree of thoroughness is appropriate for municipal or private industrial projects involving the disposal of solid wastes or wastewater into ocean waters. It also may be appropriate for navigational dredging operations, port developments, or agriculture-related management programs.

As noted in the outline in Fig. 18.4, impacts associated with effluent discharge include contaminants and suspended solids. Improvements of source control and treatment practices mitigate the potential adverse effect of the effluent. This particular assessment is further complicated by the existence of an historical sediment deposit contaminated with sulfides, metals, PCBs, and DDT. Persistent contaminants, such as DDT, from the sediment deposit are gradually being transported to the sediment surface and redistributed into the marine environment. This upward migration of contaminated sediments is caused by ocean currents and disturbances by animals. The historically deposited sediments have been partially buried by recent sediments from the wastewater outfall. Increasing treatment standards at the plants will

I. Regional Setting
  A. Physical Oceanography
  B. Historical Wastewater Discharge
  C. Receiving Water Quality
    1. Water Temperature
    2. Salinity
    3. Dissolved Oxygen
    4. Turbidity
    5. Ammonia Nitrate
    6. Metal and Other Aquatic Life Toxicants
    7. Bacteria and Viruses
  D. Sediment Quality
    1. Surface Sediments
    2. Historical Deposits of Contaminants
      a. Forces Affecting Release of Contaminants
      b. Models to Analyze the Transport and Fate of Contaminants
  E. Marine Biota
    1. Marine Habitat
    2. Plankton
    3. Kelp Beds
    4. Benthic Invertebrates
      a) Benthic Infauna
      b) Epibenthic Macroinvertebrates
    5. Demersal Fish
      a) Bioaccumulation of Toxicants
      b) Fin Erosion
    6. Pelagic Fish
    7. Coastal and Pelagic Birds
    8. Marine Mammals
    9. Rare, Threatened, and Endangered Species
  F. Areas of Special Biological Significance
  G. Uses of the Marine Environment
    1. Industrial Services Supply
    2. Ocean Dumping
    3. Navigation
    4. Water Contact Recreation
    5. Noncontact Water Recreation
    6. Ocean Commercial and Sport Fishing
    7. Shellfish Harvesting

II. Impacts and Mitigation Measures of the Plan Alternatives
  A. Criteria for Determining Significance
  B. Comparison of Alernatives
  For each alternative, the following impacts were discussed.
    1. Potential for Degradation of Marine Water Quality Resulting from Disposal of Treated Effluent
  Included effluent quality, projections, solids, and contaminant concentrations
    2. Potential for Improved Conditions for Marine Biota Resulting from Disposal of Treated Effluent
      a) Plankton
      b) Kelp Beds
      c) Benthic Invertebrates
      d) Demersal Fish
      e) Pelagic Fish
      f) Coastal and Pelagic Birds
      g) Marine Mammals
      h) Rare, Threatened, and Endangered Species
      i) Beneficial Use

Figure 18.4 Example of Contents of Environmental Impact Analysis for Sewage Treatment Facilities in a Marine Environment

reduce the release of suspended solids, normally considered an environmentally positive effect. If, however, that reduction in solids slows the burial of the historically contaminated sediment, the impact may, in fact, be adverse. This example demonstrates the sometimes extremely complicated nature of environmental impact assessment studies.

Runoff from agricultural land use is a major source of pollution within many of the nation's estuaries and bays. An example is the Susquehanna River's contribution of nitrogen and phosphorus ("nutrients") to the Chesapeake Bay. Nutrients nourish algal blooms that deprive the Bay's grasses of sunlight and deplete water of oxygen. Draining some of the most productive agricultural lands in the nation, the Susquehanna River transports mass quantities of nitrogen from fertilizer and animal waste to the Bay. Pollution reduction strategies in this case include statewide bans on detergents with phosphorus; control of runoff from urban areas, farmland, and pastures; improvements in sewage treatment; and preservation of forest and wetlands, which act as buffers to nutrient pollution inputs (U.S. Department of Interior, USGS 1995k).

Section 103 of the Marine Protection, Research, and Sanctuaries Act of 1972 applies to the transportation of dredged material for purpose of disposal in the ocean. Under the jurisdiction of the Corps of Engineers, a permit is required to determine that "the disposal will not unreasonably degrade or endanger human health, welfare, or amenities, or the marine environment, ecological systems, or economic potentialities." As with a Section 404 permit, the Environmental Protection Agency has the authority to deny the use of any defined areas as a disposal site if that will produce an unacceptable adverse effect on municipal water supplies, shellfish beds and fishery areas, wildlife, or recreational areas.

The Rivers and Harbors Act of 1899 regulates the construction of dikes, dams, bridges, and causeways across navigable waters of the United States (Sections 9 and 10); establishes harbor lines (Section 11); and grants permission for use of sea walls, bulkheads, jetties, dike, levee, wharf, pier, or other work built by the United States (Section 14).

Coastal zones and barriers are discussed in the next chapter.

18.6   Summary of Water-Related Permits and Legislation

Permits and legislation discussed in this chapter that normally may apply to evaluation of water resources impacts are summarized in Fig. 18.5 a and b.

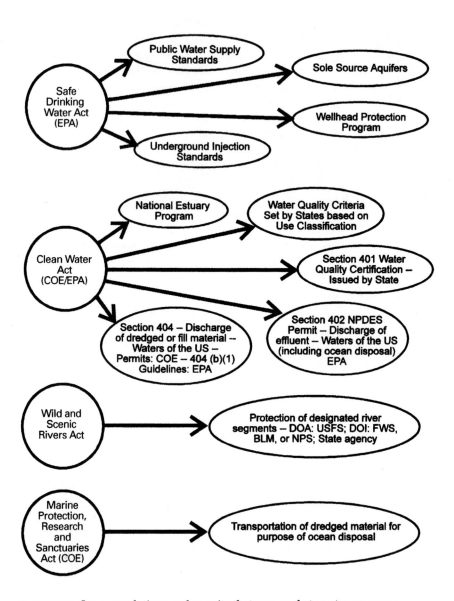

Figure 18.5a Laws, regulations, and permits that may apply to water resources.

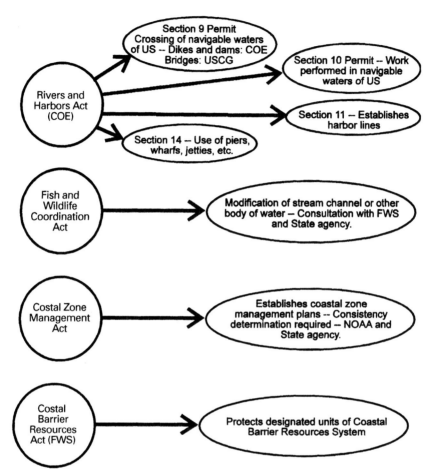

EPA = Environmental Protection Agency
Department of Defense: COE = Army Corps of Engineers
Deptartment of Interior: FWS = Fish and Wildlife Service; BLM = Bureau of Land Management; NPS = National Park Service.
Department of Commerce: NOAA = National Oceanic and Atmospheric Administration
Department of Agriculture: USFS = US Forest Service
Department of Transportation: USCG = US Coast Guard

Figure 18.5b   Laws, regulations, and permits that may apply to water resources.

19

Floodplains
and Coastal Areas

Floodplains and coastal areas are some of the most productive and valuable resources in the United States. Their functional value includes recreation, wildlife habitat, maintenance of water quality, control of life-threatening flood waters, economic development, and location of necessary infrastructure to support populations. Because of such importance, floodplains and coastal areas are protected by numerous laws and regulations. An efficient and quality environmental analysis requires a knowledge of the beneficial values, methods of impact assessment, and specific requirements of applicable statutes.

19.1 Floodplains

The assessment of possible floodplain impacts is closely related to effects on watersheds, water quality, wetlands, and coastal zones. Much of the material presented in this section of the text will also be applicable to these other related areas.

19.1.1 Legislation, regulations, and terminology

Several legislative and regulatory directives protect the use of land within floodplains. The most commonly referred to is Executive Order 11988, Protection of Floodplains. That order, signed in 1977, recognizes that "floodplains have unique and significant public values." The order requires the following:

1. All federal actions must avoid the occupancy and modification of, or the direct or indirect support of development within, the base floodplain whenever there is a practicable alternative.

2. If an action must be located on the base floodplain, agencies must minimize potential harm to people and to natural and beneficial floodplain values.

3. The public must be notified of a floodplain involvement.

As shown in Fig. 19.1, a floodplain is divided into two areas: the *regulated floodway* (which must remain unconfined or unobstructed either horizontally or vertically) and the *flood fringe* (remainder of the floodplain). The regulated floodway represents encroachment limits, so that any development permitted in the flood fringe will not result in an increase in height of more than 1 ft at any location along the stream for a 100-year flood. A *100-year flood* is a flood having a 1 percent chance of being exceeded in any given year. The 100-year flood is normally considered the "base" flood. For particularly sensitive activities, for which even a slight chance of flooding would be too great, the base flood may be the 500-year flood, which is that flood which has a 0.2 percent chance of being exceeded in any given year. The Executive Order implements and expands the Unified National Program for Floodplain Management, originally established in 1966.

For coastal and lake floodplains, a coastal high-hazard area is identified instead of a floodway. These *high-velocity zones,* or V zones, are defined as those areas which, because of tides, storm surge, topography, or other conditions, will support a 3-ft or higher wave and therefore are the areas where risk of harm or loss is greatest.

The National Flood Insurance Act of 1968, as amended, established the National Flood Plain Insurance Program. The program, administered by the Federal Emergency Management Agency (FEMA), provides for technical studies to determine the extent and frequency of flooding. Resultant mapping of all the United States is available in flood insurance studies, flood hazard boundary maps, and flood insurance rate maps for municipal and rural areas. The maps identify elevations and boundaries of the floodway, flood fringe, 100-year floodplain, and 500-year floodplain.

The National Flood Insurance Program originally applied to all coastal and riverine floodplains. The Coastal Barrier Resources Act (CBRA) of 1982 ended any federal financial assistance, specifically including flood insurance, for new construction or substantial improvements to existing structures on designated undeveloped coastal barriers. The act is further discussed in the following section.

Additionally, several federal agencies have specific flood and floodplain guidelines and regulations. For example, the U.S. Department

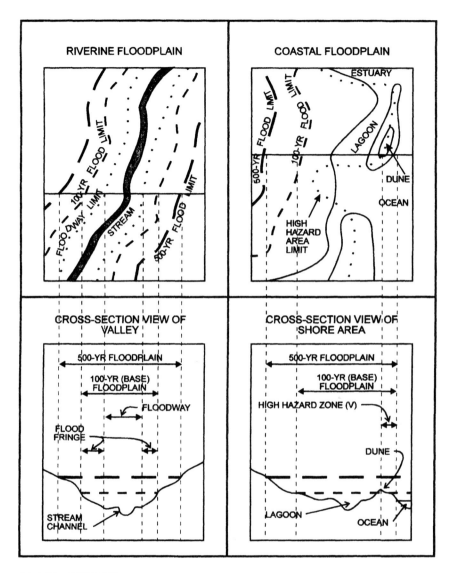

Adapted from USWRC 1978.

Figure 19.1 Floodplain glossary.

of Transportation requires hydraulic studies and floodplain impact analysis in compliance with 23 CFR Part 650. Floodplain protection and flood prevention are a significant element in the Bureau of Land Management planning system.

States and local agencies also may have related legislation and regulations. States bordering coastal areas or lakes may have individual coastal flooding provisions. Regional water basin commissions also may have implemented flood control regulations.

19.1.2 Describing existing conditions

Available mapping of limits of designated base floodplains and regulatory floodway should be obtained from FEMA. Other federal sources of information may include the U.S. Army Corps of Engineers, the U.S. Geological Survey, the U.S. Fish and Wildlife Service, and the Department of Commerce National Atmospheric and Oceanic Administration. State and local agencies should be contacted to identify applicable regulations, codes, or zoning provisions. Flood or floodplain management plans should be reviewed to identify permitted activities and land uses within the base floodplain.

To assess possible cumulative impacts, it is necessary to obtain information on other planned projects or actions within the watershed(s) and to review the characteristics of, and the environmental documents prepared for, these other projects.

As dictated by the expected magnitude of impact and the scoping process, detailed information on flood water flow characteristics should be reviewed from appropriate flood insurance studies or hydraulic studies.

If it appears that the proposed project or action will encroach on a base floodplain, it is necessary to define the natural and beneficial floodplain values and functions. Functional value determinations are discussed in greater detail in Chap. 20.

Floodplains, and wetlands in floodplains, have significant natural and beneficial values including

- Natural flood storage and conveyance
- Improvement of water quality by filtering sediment and contaminants
- Groundwater recharge
- Habitat and critical energy source for large and diverse populations of plants and animals
- Cultural resources, such as archaeological sites, parks, and aesthetic recreational areas

In summary, the environmental impact analyst must have an understanding of previous studies; designated floodplain limits; existing use and values; local goals and objectives for floodplain use and development; and other planned projects and actions in the project area. The level of detail of acquired information and data will be guided by the expected magnitude of impact and the results of scoping with state and local agencies and the public.

19.1.3 Impact analysis

The assessment of potential floodplain impacts progresses systematically through several steps. First, it must be determined whether the project will cause an encroachment of the floodplain or floodway. The process then continues through development and feasibility assessment of avoidance alternatives, determination of risk to human life and property, and assessment of impacts to functional values.

Encroachment determination. The first step is to determine whether the proposed project or action will cause an encroachment on the base floodplain, either directly or indirectly through induced secondary development. If floodplain issues are identified as an area of concern, the multidisciplinary team will most likely include engineers evaluating changes in the hydraulic and hydrologic characteristics of the study area streams and floodplains due to the proposed project or action. These studies will identify encroachments in the base floodplain or regulatory floodway; calculate changes in flood levels, runoff quantity, and velocity; identify expected groundwater interception or flow changes; and assess flood water flow characteristics at specific, possibly problem, areas such as culverts and ponding potential sites for determination of localized flood water flow attributes. It is often relevant to potential impacts to classify the encroachment as transverse or longitudinal. Longitudinal encroachments are of much greater concern.

The other aspect of identifying applicability of Executive Order 11988 is to determine whether the proposed project or action will cause the direct or indirect support of development within the base floodplain. Although the modification of floodplains most clearly results from actions located in the floodplain, it can also result from actions out of the floodplain. The location of major developments and activities outside the floodplain may support associated, subsequent, secondary development within the floodplain. Floodplain development can be indirectly supported by providing infrastructure, such as water and wastewater systems, power supplies, highway and secondary road networks, mass transit systems, and airports outside the floodplain.

Avoidance alternatives. The next step in the analysis process, if an encroachment is identified, is to develop and evaluate alternatives to avoid the encroachment. No floodplain encroachment is permitted unless there is no practicable alternative. The avoidance alternative concept also applies to identified adverse effects. The first and preferred option is to avoid encroachment or adverse effects.

The analysis must include actual development of real avoidance alternatives to a level of detail to permit analysis of practicability. Considered alternatives must include alternative sites, alternative actions, and the no-action alternative. Alternative actions are those that may substitute for the proposed action by constituting new solutions or approaches which serve the same function or purpose as that proposed, but which have less potential for harm. The identified purpose and need of the proposed project or action will direct development of evaluation criteria for determining practicability. Other criteria may include legal considerations, costs, and natural, social, and economic impacts.

Risk to life and property. If avoidance of a floodplain encroachment is not practicable, the analysis continues with the evaluation of potential risks to life and property from flood hazards. Risk assessment examines

- Calculation of changes in flood elevations and flow characteristics
- Whether the infringement is located within the regulated floodway or the flood fringe
- Potential destructive velocity flows, flood-related erosion, mudslides, sinkholes, etc.
- Possible combination of flood sources which may flood simultaneously

Natural and beneficial floodplain values. The Unified National Program for Floodplain Management (Federal Emergency Management Agency 1986) gives direction for analysis of impacts to natural beneficial floodplain values. The assessment should include potential direct, indirect, cumulative, short-term, and long-term effects. Considered effects should generally include

- Accelerated runoff, with resultant erosion and an upset of the balance of erosion and deposition within the stream
- Increased transport and loading of chemical and organic contaminants
- Increase in flood peaks due to accelerated runoff reducing the amount of water entering the ground
- Decrease in groundwater recharge
- Blocked or diverted groundwater flow
- Changed distribution of sediment on natural shorelines
- Removal of shelter, feeding, and nursery areas for wildlife and fishery resources

The types of potential impacts assessed are those also associated with coastal areas, water quality, wetlands, and vegetation and wildlife resources. The assessment, as with these related areas of potential impacts, should be based on an understanding of ecological principles and the "ripple" effects that can occur within a balanced ecosystem when a major change from human intervention, such as development within a floodplain or wetland, takes place.

19.1.4 Mitigation

Executive Order 11988 requires measures to minimize, restore, and preserve if the proposed project or action causes harm to lives and property or to natural and beneficial floodplain values. Minimization applies to harm to lives and property. Possible measures include using structures to cross floodplains instead of fills, providing adequate flow circulation, reducing grading requirements, and preserving free, natural drainage when possible.

Any encroachment within a regulatory floodway must undergo a *consistency analysis* to determine if the proposed project or action is consistent with, or requires revision to, the regulatory floodway. Coordination is required with the Federal Emergency Management Agency and appropriate state and local agencies. Revisions may include such measures as construction of dams and reservoirs; dikes, levees, or floodwalls; channel alterations; or high-flow diversions and spillways. If the conveyance and storage capacity of the floodplain is reduced, measures to compensate and provide for equal conveyance and storage must be assessed. If revisions to the regulatory floodway are required, such revisions must be acceptable to the Federal Emergency Management Agency and to state and local agencies.

The restoration and preservation directives of Executive Order 11988 apply to natural and beneficial floodplain values. The proposed project or action must incorporate all possible measures to preserve and restore floodplain values. Numerous specific measures are suggested in the Executive Order 11988 implementation guidelines (NWRC 1978) and the floodplain management program (Federal Emergency Management Agency 1986). These are examples of suggested measures related to water quality, groundwater recharge, and aquatic and wetland ecosystems:

- Maintain wetland and floodplain vegetation buffers to reduce sedimentation and delivery of chemical pollutants to the water body.
- Control agricultural activities to minimize nutrient inflow.
- Control urban runoff and other storm water and point and nonpoint discharges.
- Control methods used for grading, filling, soil removal, and soil replacement to minimize erosion and sedimentation during construction.

- Prohibit the location of potential pathogenic and toxic sources on the floodplain.
- Require the use of pervious surfaces wherever possible.
- Design construction projects for water retention.
- Dispose of spoils and waste material so as not to contaminate ground or surface water or change land contours.
- Identify and protect wildlife habitat and other vital ecologically sensitive areas from disruption.
- Require topsoil protection programs during construction.
- Control wetland drainage, channelization, and water withdrawal; reestablish damaged floodplain ecosystems.
- Minimize tree cutting and other vegetation removal.
- Design floodgates and seawalls to allow natural tidal activity and estuarine flow.

Many of the above measures deal with overall floodplain and watershed management as opposed to possible project-specific techniques. The principles and objectives set forth apply to sound management practices for minimizing impacts, not only to floodplains, but also to watersheds, coastal areas, surface waters, groundwater, wetlands, and biotic communities. Project- or action-specific techniques will be developed based upon these principles.

After identification of impacts, steps necessary to minimize impacts, and opportunities to restore and preserve floodplain values, the alternatives are reevaluated for feasibility of limiting the proposed project or action or taking no action.

19.1.5 Public notification

Regulations require early notification to the public of potential floodplain impacts of proposed projects or actions. The public input process should continue throughout the development and evaluation of avoidance alternatives, assessment of impacts, and design of mitigation measures, if any are required. Normally this notification is accomplished as part of the NEPA process through scoping, public meetings, and agency consultations. The Environmental Assessment or Draft Environmental Impact Statement will compare proposed alternatives for impact on floodplains. If a floodplain encroachment is required by the selected alternative, the Finding of No Significant Impact or Final Environmental Impact Statement must contain a *floodplain finding,* explaining to the public (1) why the proposed action must be located in the floodplain, (2) why the considered avoidance alternatives are not practicable, and (3) whether the action conforms to applicable state or local floodplain protection standards.

The statement of findings (including explanatory information) must be issued for all projects or actions proposed within or impacting the floodplain, including proposed actions whose impacts are not significant enough or are not otherwise required to complete an Environmental Impact Statement.

19.1.6 Summary of the process

These are the basic steps in compliance with Executive Order 11988:

1. Gather information on floodplains, state and local plans, other planned projects, and natural and beneficial floodplain values.

2. Consult with FEMA and state and local agencies.

3. Determine whether the proposed project or action involves a direct encroachment and/or encroachment through support of floodplain development.

4. If there is an encroachment, notify the public for input and develop and evaluate avoidance alternatives for practicability.

5. Determine risk to life and property and the effects on natural and beneficial floodplain values.

6. Develop mitigation measures for adverse effects.

7. Reevaluate alternatives.

8. If the project proceeds, notify public with an explanatory finding.

19.2 Coastal Zones and Barriers

Laws and regulations protecting the coastal areas of the United States have developed because of a very basic fact which should be understood by the environmental impact analyst. That fact is that there is intrinsic conflict between the numerous "required" uses of coastal areas by federal and nonfederal programs (Fig. 19.2).

19.2.1 Federal programs affecting coastal areas

Federal agencies are charged with administering programs with missions as diverse as the uses of the coast. Often these programs result in nationwide conflicts between development and conservation. For example:

- The Federal Emergency Management Agency spends billions of dollars through its Federal Insurance Administration (FIA).

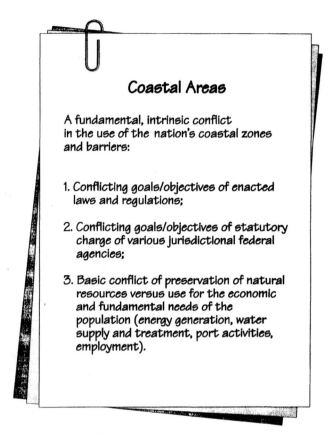

Figure 19.2 Coastal areas.

- The Army Corps of Engineers is responsible for port maintenance and construction for the nation's seaports and for navigation, flood control, and erosion control projects in coastal states.

- The Environmental Protection Agency provides assistance through grants for wastewater treatment facilities in coastal areas.

- The Federal Highway Administration (FHWA) provides assistance for construction of roads and bridges.

- The U.S. Coast Guard issues permits for construction of bridges and causeways across navigable waters.

- Numerous federal programs are available to assist in postdisaster reconstruction (many authorized by the Disaster Relief Act of 1972), even if that reconstruction encourages continued incompatible development within high-hazard coastal areas.

Many of these federal assistance programs have facilitated development and population growth in hazardous and vulnerable coastal areas.

19.2.2 Coastal Barriers Resources Act

The purpose of the Coastal Barrier Resources Act (CBRA) is to restrict growth-inducing federal expenditures and financial assistance
in remaining coastal barrier areas. The act establishes the Coastal
Barrier Resources System (CBRS), composed of undeveloped coastal
barriers along the Atlantic and Gulf coasts that are depicted on a set
of maps. The program is under the jurisdiction of the Department of
Interior, Fish and Wildlife Service.

Any proposed project or action alternative which may impact a designated unit of the CBRS should be evaluated, in consultation with
the U.S. Fish and Wildlife Service, for potential direct and indirect effects. Impacts should be comparably quantified, by alternative, and
the results of the U.S. Fish and Wildlife Service consultation should
be documented in the draft environmental document. The final environmental document should indicate resolution of any issues pertaining to resources protected by the act. If the proposed project or action
is not consistent with the purposes of the act, it cannot receive federal
assistance unless it is determined to be a proper exception.

19.2.3 Coastal Zone Management Act

The Coastal Zone Management Act of 1972, as amended in 1990, provides for establishment of coastal zone management plans (CZMPs)
in coastal states. The act is under the jurisdiction of the Department
of Commerce, National Oceanic and Atmospheric Administration
(NOAA), with joint concurrence with the Environmental Protection
Agency. Among other things, the CZMPs include

- An identification of the inward and seaward boundaries of the
  coastal zone subject to the management program

- A definition of what constitutes permissible land uses and water
  uses within the coastal zone which have a direct and significant impact on the coastal waters

- An inventory and designation of areas of particular concern within
  the coastal zone, such as natural areas, wildlife habitat, and ports

- Broad guidelines on priorities of uses in particular areas, including
  specifically those uses of lowest priority

- A definition of the term *beach* and a planning process for the protection of, and access to, public beaches and other public coastal areas
  of environmental, recreational, historical, aesthetic, ecological, or
  cultural value

- A planning process for energy facilities likely to be located within or to significantly affect, the coastal zone
- A planning process for assessing the effects and control of shoreline erosion

The plan also must include assurances that local regulations within the coastal zone do not unreasonably restrict or exclude land and water uses of regional or national significance, such as energy facilities. This provision again points to the inherent conflict between various programs vying for use of coastal areas. These needs must be considered in the development of a CZMP and in an environmental impact analysis of such a plan. Possible effects of management plans may be direct, secondary, beneficial, or adverse.

For example, among the most important economic activities affected by coastal management decisions are the nation's seaports. Waterborne commerce contributes significantly to local, state, regional, and national economic growth. As the demand for deep-draft ports increases, associated impacts of dredging to accommodate those vessels also increase. Many issues surrounding port development and expansion also affect the general economic growth in coastal areas.

The assessment of possible secondary and cumulative effects of actions within coastal areas also can be critical, especially for infrastructure improvements which provide access, water supply and wastewater treatment facilities, navigation, flood control, and erosion control. For example, construction of a large, regional wastewater treatment facility may make possible the development of housing projects in areas previously served by small municipal systems. This secondary development may, in turn, destroy wetlands, increase demand for better highways, increase air pollution, and increase the demand for flood protection.

19.2.4 Coastal Non-point Pollution
Control Program

The 1990 amendments of the Coastal Zone Management Act established the Coastal Non-point Pollution Control Program. Each state's coastal zone management plan must contain enforceable policies and mechanisms to implement the Non-point Pollution Control Program. The Coastal Non-point Pollution Control Program is an important source of information to describe existing conditions in a potential project area. It identifies land uses which individually or cumulatively may cause or contribute significantly to the degradation of coastal waters and those areas of coastal waters threatened by reasonably foreseeable increases of pollution loadings from new or expanding sources. It also provides guidance on specific measures to manage sources of nonpoint pollution.

19.2.5 Impact analysis consultation

All alternatives of the proposed project or action should be reviewed for location within an area included in an approved CZMP. If such involvement is apparent, coordination should begin immediately with the state coastal zone management agency, the U.S. Fish and Wildlife Service, and other federal and state agencies as applicable. The exact changes in or adjacent to the designated coastal zone area predicted to occur with each proposed alternative should be quantified to permit evaluation of potential impacts.

19.2.6 Consistency determinations

The Coastal Zone Management Act requires that all federal agency activities within or outside the coastal zone that affect any land or water use or natural resource of the coastal zone be consistent to the maximum extent practicable with the enforceable policies of the approved CZMP. Initial early consultation should be conducted with the appropriate state agency to identify the provisions of the CZMP which are related to the proposed activity and the information that will be necessary to determine consistency. A consistency determination must contain

- A brief statement indicating whether the proposed project or action will be undertaken in a manner consistent to the maximum extent practicable with the enforceable policies of the management plan
- An evaluation or the relevant provisions of the management plan that forms the basis for the statement
- A detailed description of the activity, its associated facilities, and supporting information

The amount of detail in the statement evaluation, activity description, and supporting information should be commensurate with the expected effects of the proposed project or action on the coastal zone.

The consistency determination should be prepared and sent to the appropriate state agency at the earliest possible time during environmental studies, but it must be transmitted at least 90 days prior to any final decisions by the sponsoring agency. The state agency has 45 days to respond.

If the proposed project or action is not a federal action, but is a federally assisted state or local action, then the state CZMP agency will be notified as part of the normal state clearinghouse notification process. The state agency will issue an opinion on consistency.

The draft environmental document should document coordination with the state agency and describe the potential impacts of all proposed alternatives. The final environmental document should document the state agency's agreement on consistency with the state CZMP. Projects or activities found not to be consistent with the plan, or for which there is disagreement between state and federal agencies, enter a mediation process. If the process determines the proposed project or action is not consistent with the CZMP, the project cannot proceed with federal funding or assistance.

Federally licensed and permitted activities also require consistency with the state CZMP. The application for the federal permit or license must contain a consistency certification.

19.2.7 National Estuary Program

The National Estuary Program was established to identify nationally significant estuaries, protect estuary water quality, and enhance estuary living resources. The program was created in 1987 under the Clean Water Act and is administered by the Environmental Protection Agency.

Management conferences are held for estuaries in the program. The purpose of the conferences is to identify priority problems in the estuary and develop comprehensive conservation and management plans (CCMPs) to address the problems. Appropriate management action and mitigation measure plans are then developed to identify specifically who will carry out the action, when and how the action will be conducted, and how the action will be financed. The National Oceanic and Atmospheric Administration and the U.S. Fish and Wildlife Service are usually represented on the committee for each estuary program to ensure consistency with the Coastal Zone Management Act and the Coastal Barriers Resources Act.

The estuary programs have the responsibility to review proposed federal development projects for consistency with the comprehensive conservation and management plan.

20

Wetlands

Wetlands are one of the most productive ecosystems on earth. Because wetlands can be located within floodplains, provide important water quality functions, and provide valuable wildlife habitat, this chapter is closely related to those chapters of the text dealing with water resources, floodplains, and vegetation and wildlife. Information contained in those other chapters will not be repeated here; the reader is encouraged to combine those chapters with this one to gain a complete understanding of the process of evaluating potential wetland impacts and mitigation measures.

Several steps are involved in the identification and assessment of wetland impacts. The four major federal agencies involved are the Environmental Protection Agency (EPA), U.S. Fish and Wildlife Service (FWS), U.S. Army Corps of Engineers (Corps), and Natural Resources Conservation Service (NRCS). Coordination related to wetlands also should include the state fish and wildlife agency (agencies) and the environmental resources agency.

20.1 Legislation and Regulations

Executive Order 11990 is the primary regulation protecting wetlands of the United States. The Clean Water Act also provides for protection of wetlands, particularly within the Section 404 permit process and the Section 404(b)(1) guidelines. The Food Security Act (Swampbuster) applies to wetlands on agricultural lands.

Because of the numerous regulatory requirements, wetlands identification, assessment, and permitting became a complex task in the 1980s. By the middle 1990s, several major steps had been taken to avoid duplication, clarify definitions, reduce time for permitting, and ensure consistency among the various regulatory guidelines and manuals. These were some of the improvements:

- Permit flexibility for small landowners
- A nationwide permit for single-family housing
- Less vigorous permit review for small projects having minor environmental impacts or affecting wetlands of minimal functional value
- A streamlined permit process
- Better analytical methodologies for wetland functional assessments
- Consistency in wetland delineation techniques
- Incentives for private participation in the Wetland Reserve Program of the NRCS
- Clarification in definition of waters of the United States, previously converted cropland, and artificial wetlands
- Endorsement of the use of mitigation banking under Section 404 regulatory program
- Promotion of voluntary incentives for wetland restoration on private lands

Because of the numerous overlapping regulations and the ongoing changes, it is important that qualified personnel assessing wetlands impacts be up to date on the latest guidelines and regulations.

20.2   Data Collection

Steps to describe existing wetland resources are summarized in Fig. 20.1. A good place to begin is with a check of the National Wetland Inventory (NWI) mapping for the site or corridor of the project. NWI maps are available from the U.S. Fish and Wildlife Service. These nationwide maps were prepared mostly by viewing aerial photography and always should be verified in the field.

Coordination and consultation with the EPA, FWS, Corps, and particularly state or regional agencies will likely yield information on existing wetland resources in a particular area. Such coordination, if conducted as part of the scoping and early coordination effort, also can identify study requirements and methodologies.

For example, for a project in Maryland (U.S. Department of Transportation, FHWA, and Maryland State Highway Administration 1988), coordination and request for information from the Water Resources Administration of the Department of Natural Resources (Waterways Permit Division) yielded a general outline of the types of

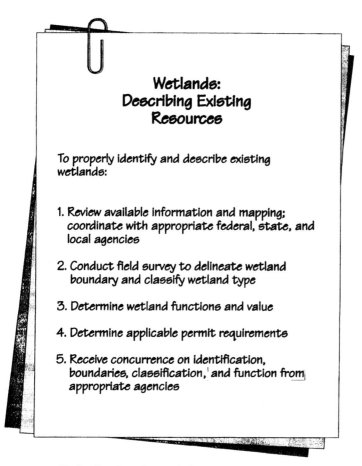

Wetlands:
Describing Existing
Resources

To properly identify and describe existing
wetlands:

1. Review available information and mapping;
   coordinate with appropriate federal, state, and
   local agencies

2. Conduct field survey to delineate wetland
   boundary and classify wetland type

3. Determine wetland functions and value

4. Determine applicable permit requirements

5. Receive concurrence on identification,
   boundaries, classification, and function from
   appropriate agencies

Figure 20.1   Wetlands—describing existing resources.

nontidal wetlands that presently exist in the project area and results
of a preliminary analysis of the NWI maps. Wetland listings were then
provided by USGS Quad, including classification, such as

- *PF01A*—Palustrine forested, temporarily flooded, broad-leaved deciduous vegetation
- *R2OWH*—Riverine, lower perennial, open water, permanently flooded
- *PEM5A*—Palustrine, emergent, temporarily flooded, narrow-leaved
  persistent vegetation
- *PSS1A*—Palustrine, scrub and shrub, broad-leaved deciduous, temporarily flooded

The Coastal Resources Division recommended that the following information be covered in the environmental analysis:

- Field-identified data on the vegetative species including dominant, understory, and herbaceous plant types
- Soil characteristics of the wetlands, including hydrologic regime (for example, temporary, saturated, seasonal, permanent) and drainage class (for example, poorly drained, very poorly drained)
- Wetlands acreage impacted, by type
- Aquatic and terrestrial wildlife in the project area
- Benthic invertebrates inhabiting the streams or rivers
- Details of proposed mitigation for wetland impacts
- Wetland boundary delineation performed in the field and flagged with bright, plastic ribbon and provided on map of the project

As noted by the above example, federal, state, and local agencies can be important sources of information on existing wetland resources and on requirements for subsequent studies of wetland impacts. Other important sources of information may be environmental documents prepared for other projects or actions in the immediate area, or management and land-use plans. If it appears that a project will impact a wetland, all information obtained from existing sources should be verified in the field.

20.3   Field Surveys and Delineations

Wetlands are defined and classified by the presence of three factors: wetland hydrology, hydric soils, and hydrophytic vegetation (Fig. 20.2). *Hydrology* refers to the water characteristics of the area, such as the amount of time an area is wet or flooded a year, or intermittence* of streams. Wetland hydrology exists when areas are permanently inundated or saturated to the surface at some period during the growing season. *Hydric soils* are those that are wet long enough to periodically produce anaerobic (lacking free oxygen) conditions, thereby influencing the growth of plants. *Hydrophytes* are plants growing in water or on a substrate that is at least periodically deficient in oxygen as a result of excessive water content (U.S. Department of Interior, Fish and Wildlife Service 1979).

---

*An intermittent stream does not have a continuous flow and is dry part of the time.

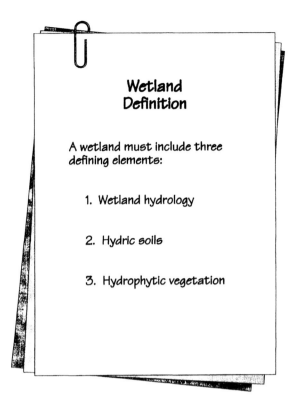

Figure 20.2   Wetland definition.

Although the federal agencies with jurisdiction over, and interest in, wetlands used to have individual methodologies for identification and delineation of wetlands, the EPA, FWS, Corps, and the NRCS now use a methodology jointly acceptable by all. Whereas previous individual methodologies emphasized one or more of the three factors, the accepted methodology now is to use vegetation, soils, and hydrology to define and delineate wetlands.

The acceptable methodology, however, has been, and may continue to be, refined, including revisions to definitions of wetlands. For example, a joint-agency identification and delineation methodology prepared in 1989 was replaced with a new manual in 1990. Both were then rescinded in 1991, and all agencies went back to using the previous Corps Delineation Manual of 1987. Previous inconsistencies in provisions of the Clean Water Act Section 404(b)(1) guidelines and the Food Security Act (Swampbuster) Manual have been eliminated, especially regarding classification of previously converted cropland and artificial wetlands.

It is therefore important that wetland study team members be experienced professionals with up-to-date information on the most recent guidance. Because of an interest within the profession to improve the quality and consistency of wetland delineations, and to streamline the regulatory process, the Corps, as authorized by the Water Resources Development Act of 1990, has developed a program for the training and certification of individuals as wetland delineators. Wetland delineations submitted to the Corps for permit applications which have been prepared by certified wetland delineators will be given expedited Corps review.

The next step in the process of identifying wetlands is the preparation for a field view of the project area. Wetlands are defined as "areas that are inundated or saturated by surface or groundwater at a frequency and duration sufficient to support, and that under normal circumstances do support, a prevalence of vegetation typically adapted for life in saturated soil conditions" (U.S. Department of Defense, Army Corps of Engineers, 1987).

The NRCS soil survey should be reviewed. Each NRCS local office will have a list of soil units classified as hydric. Review the soil maps carefully, and note the location of any hydric soils. A list of vegetation classified as hydrophytic should be available for the region of the proposed project from either a state agency or the U.S. Fish and Wildlife Service. The FWS publishes a national list of plant species that occur in wetlands (U.S. Department of Interior, FWS 1988).

Vegetation species are classified into indicator groups as obligate, facultative, or upland, based on the expected occurrence of the species in wetlands (Fig. 20.3). *Obligate* species are generally found only in wetlands. *Facultative* means the species is a border plant that can occur both inside and outside wetlands. *Upland* species are normally not found in wetlands.

Because the definition of wetlands requires a prevalence of hydrophytic vegetation, the term *prevalence* must be defined. If a visual observation estimate indicates that obligate and facultative wet plant species cover a greater percentage of the area than facultative upland and upland plants, then the area is determined to have a prevalence of hydrophytic vegetation. If not, the area is a nonwetland.

If visual observation cannot determine prevalence, a formula must be used based on transect techniques of observation. For the purposes of the formula, an ecological index has been assigned to each indicator group, from 1 for obligate to 5 for upland. The prevalence index is then calculated based on the frequency of occurrence of each indicator species group and the ecological indices. Results of the formula indicate whether a prevalence of hydrophytic vegetation meeting the definition of wetland exists.

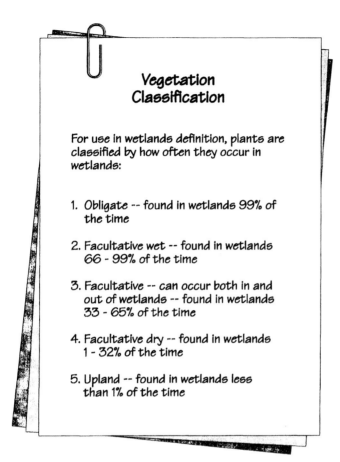

Vegetation
Classification

For use in wetlands definition, plants are
classified by how often they occur in
wetlands:

1. Obligate -- found in wetlands 99% of
   the time

2. Facultative wet -- found in wetlands
   66 - 99% of the time

3. Facultative -- can occur both in and
   out of wetlands -- found in wetlands
   33 - 65% of the time

4. Facultative dry -- found in wetlands
   1 - 32% of the time

5. Upland -- found in wetlands less
   than 1% of the time

Figure 20.3 Vegetation classification.

Field views for the presence of wetlands can be general or very specific. If a formal flagging of wetlands boundaries is required, flags are
placed in the field to delineate the boundary of the wetland. This delineation uses the method established by the four agencies and requires identification and classification of plant species, testing of soil,
and site observation and further study of local hydrologic conditions.

Each wetland is classified according to type (Fig. 20.4). The U.S.
Fish and Wildlife Service classification system is the one most commonly used. The system is too extensive to explain in detail in this
text; nor would it be consistent with the purpose of this text. In general, however, the wetlands are first classified by system, including marine, estuarine, riverine, lacustrine (lake), or palustrine (border areas
between the other systems and uplands). The next level of classification for the first four systems is the subsystem, such as tidal and in-

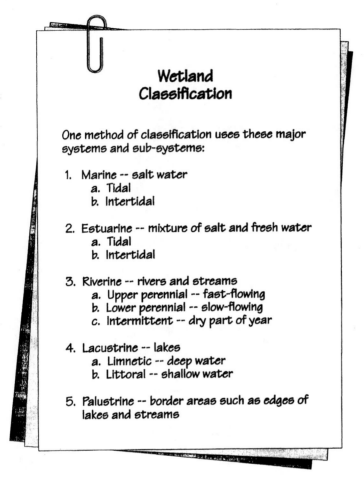

Wetland Classification

One method of classification uses these major systems and sub-systems:

1. Marine -- salt water
   a. Tidal
   b. Intertidal

2. Estuarine -- mixture of salt and fresh water
   a. Tidal
   b. Intertidal

3. Riverine -- rivers and streams
   a. Upper perennial -- fast-flowing
   b. Lower perennial -- slow-flowing
   c. Intermittent -- dry part of year

4. Lacustrine -- lakes
   a. Limnetic -- deep water
   b. Littoral -- shallow water

5. Palustrine -- border areas such as edges of lakes and streams

Figure 20.4 Wetland classification.

tertidal for marine and estuarine; upper perennial (fast-flowing), lower perennial (slow-flowing), or intermittent for riverine; or limnetic (deep water) or littoral (shallow water) for lakes. The palustrine system does not have subsystems. The classification continues with descriptors for bottom types and vegetation types (such as emergent or persistent) and flooding descriptors.

20.4   Functional Value Assessment

A functional assessment also is made during the field survey. Several methodologies have been used for functional assessment (U.S. Department of Defense, Army Corps of Engineers 1979; U.S. Department of Transportation, FHWA 1982; and U.S. Department of Defense, Army Corps of Engineers, U.S. Department of Transportation,

FHWA 1987), and the techniques continue to improve and become more consistent. Regardless of the methodology used, the following features of the wetland are important to consider when the functional value of wetlands is defined:

- Groundwater recharge value
- Water quality and sedimentation value
- Wildlife habitat
- Flood protection
- Biological diversity
- Recreation
- Aesthetics

Other factors entering into the evaluation of wetland value is its uniqueness and its relationship to other wetlands in the area or region.

Although all four agencies may agree on the methodology for identification of wetlands, there often will be differences in jurisdictional interest. The Corps, for example, has jurisdiction in wetlands that qualify for classification under their 404 permit. Therefore, the description of wetlands in the study area may need to designate which of the identified wetlands would fall under Corps 404 jurisdiction as "waters of the United States." Some surface waterways will qualify for 404 jurisdiction as waters of the United States, but may not meet the criteria to be designated as *wetlands*.

20.5 Interagency Concurrence on Existing Resources

Results of the field view should be circulated to agencies having jurisdiction and interest over wetlands for the project site. If a detailed wetland delineation has been conducted, a separate report should be prepared to document the study, including mapping of delineated wetlands and information on classification and functional value. Agencies should be invited to a field view to concur with the boundaries, classification, and functional value. These same agencies then are later contacted to review the results of the impact assessment and proposed mitigation measures.

For example, a field view of a delineated and flagged wetland may include representatives of the EPA, Corps, FWS, and state agencies. For each identified and flagged wetland, concurrence should be requested on

- Whether marginal areas are indeed wetlands (such as, it was agreed that no wetland existed at this site outside of the stream channel)
- Wetland classification

- Wetland boundaries
- Wetland functional values
- Estimated affected acreage—to be refined based on results of field view
- Whether specific identified wetlands are outside the area of potential impact and are eliminated from further consideration
- Whether mitigation will be required

If only minor involvement with wetlands is discovered, the information sent to agencies can be in the form of the draft text to appear in the DEIS and can include the results of impact analysis and proposed mitigation. In all cases, however, the EPA, FWS, Corps, and appropriate state agencies should be contacted for consultation on concurrence with both the identification and impacts assessment at the DEIS stage. In fact, the DEIS must show proof of this consultation. Agreement on specific mitigation can wait until the FEIS because it will be specific to the preferred alternative.

20.6 Impact Assessment

The most common impact to wetlands is from filling or draining to make land available for other uses. To begin evaluation of the impact of the proposed project alternatives, the physical characteristics of the project should be reviewed to ascertain if any part of the identified wetlands will be physically destroyed. The acres of wetlands affected, by wetland type, should be calculated for each proposed alternative.

The EPA Section 404(b)(1) guidelines (U.S. Environmental Protection Agency 1980), discussed in Chap. 18, lists types of impacts to be considered for aquatic ecosystems, including wetlands. Effects may include changes in water levels, flow characteristics and circulation patterns, or flooding frequencies. Changes in bottom types and other substrate conditions will affect the ability of the wetland to sustain vegetation and wildlife populations. Creation of increased runoff may introduce contaminants or toxins to the wetland ecosystem or may increase nutrient loadings that produce algal blooms and reduce oxygen in the water. Changes in current patterns or velocities can affect coastal and riverine wetland ability to flush contaminants or filter sediments.

The assessment must also consider secondary effects and long-term effects. If a project will change the quality or volume of water flow or drainage patterns, wetland characteristics will be changed. Many projects located outside wetlands may induce additional development. The development, with required infrastructure improvements, could then cause secondary effects and additional stress on wetlands.

Finally, the analysis must consider cumulative effects. The analysis should identify other projects within the area of influence of the wetland and predict the integral increase in impact caused by the proposed project or action. The integral increase in impact, or degradation of wetland functions or values, may constitute an adverse impact even if the project, by itself, would not produce adverse effects. The assessment should consider the relationship of the particular wetland area affected to the availability of similar resources in the project area.

20.7 Mitigation

The Section 404(b)(1) guidelines establish a three-step sequence for mitigating potential adverse impacts to the aquatic environment associated with a proposed discharge of dredged or fill material (Fig. 20.5). The three steps are avoidance, minimization, and compensation for unavoidable impacts.

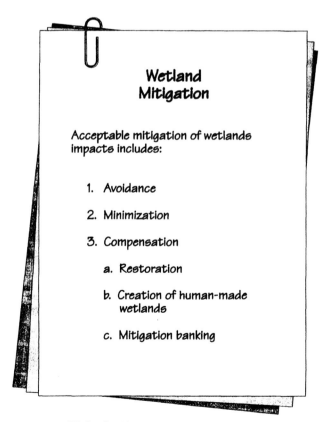

Figure 20.5 Wetland mitigation.

20.7.1 Avoidance alternatives

If a proposed project or action will involve the filling of wetlands, the first step in mitigation is development and evaluation of alternatives to avoid the wetlands. In some cases this may be a very simple and obvious task, to demonstrate that there are no practicable alternatives to the loss of wetlands. In other cases, these alternatives should be developed to a level of detail equal to that of the other proposed alternatives and should be thoroughly evaluated for feasibility and practicability. Alternatives should include actions as well as sites or locations that may meet the identified project purpose and need.

The Clean Water Act prohibits discharge of dredged or fill material if there is a practicable alternative to the proposed discharge which would have a less adverse impact on the aquatic ecosystem, if the avoidance alternative does not have other significant adverse environmental consequences. In other words, if an alternative avoids a wetland impact or has less impact than the original proposed alternatives, there is basically no decision remaining to be made. The avoidance, or less-impact, alternative must be chosen.

A significant feature of the improvements in the 1990s (Interagency Working Group on Federal Wetlands Policy 1995) in the regulatory process was the introduction of flexibility into the required analysis of avoidance alternatives. The Section 404(b)(1) guidelines are the standards by which all Section 404 permit applications are evaluated. The changes in the guidelines clarified that the level of review of project alternatives should be commensurate with the severity of the impact. The new guidance specifically stated that for projects causing only minor or negligible impacts, a detailed analysis of avoidance alternatives should not be conducted. Minor impacts are defined as having little potential to degrade wetlands, such as

- Impacts that occur in wetlands of limited functional value
- Impacts of less than an acre
- Impacts with little potential for secondary or cumulative effects
- Impacts expected to be temporary (less than a year)

The revised guidelines also recognized that avoidance alternatives may not produce much difference in the degree of impact. Therefore, "an elaborate search for and analysis of practicable alternatives is *not* required if it is reasonably anticipated that the difference between the environmental impacts of the proposed project and potential alternatives would be only minimal."

The assessment of feasibility and impacts of avoidance alternatives should be completed to be included in the DEIS or EA stage of the environmental process.

20.7.2 Minimization

If avoidance is not possible, the analyst should evaluate all possible measures to minimize harm to the wetlands and its functional values. As discussed in Chap. 18 and shown in Fig. 18.2, the Section 404(b)(1) guidelines in 40 CFR Part 230, Subpart H, list possible applicable mitigation actions in eight general areas of impact. More than 40 specific measures are suggested for use.

Although a detailed analysis may wait until preparation of the final environmental document, a preliminary evaluation of the feasibility and success of mitigation measures can be done in a time frame to be included within the DEIS. Often if a project has many alternatives, with differing wetland impacts, available mitigation will be generally suggested within the DEIS to enable the reader to understand the degree to which the impact can be successfully mitigated or avoided. The commitment to a detailed plan for mitigation is then included within the FEIS for the selected alternative.

20.7.3 Compensation

The third means of mitigating wetland impacts is compensation, including restoration of degraded wetlands or construction of replacement wetlands for those lost. Compensatory actions should be considered only after all appropriate and practicable minimization has been employed and adverse impacts remain. Restoration is preferred over creation of new wetlands because of the ever-present uncertainty of the success of created wetlands. On-site, in-kind compensatory mitigation is preferred to off-site, out-of-kind.

Restoration. An example of restoration mitigation is a project where construction would result in removal of riparian* scrub habitat (County Sanitation Districts of Los Angeles County 1994). Mitigation was to restore riparian scrub and forest habitat at sites on district land that presently support ruderal† and grassland vegetation. At least 2 acres of riparian scrub habitat would be restored for each acre removed (2:1 ratio). A specific riparian habitat restoration plan was prepared and implemented that included the following elements:

- Planting locally native riparian trees and shrubs collected from local genetic stock

---

Riparian describes vegetation on the bank of a natural watercourse, such as a river or lake.

†*Ruderal* vegetation refers to weeds and other plants that grow where the natural vegetational cover has been disturbed by humans.

- Implementing necessary irrigation, weed control, herbivore control, and other cultivation measures for tree and shrub plantings
- Establishing habitat restoration success criteria based on achieving native vegetative cover and diversity and wildlife habitat equal to or greater than that of habitat that is removed
- Initiating vegetation restoration and monitoring prior to removal of vegetation
- Monitoring the success of habitat restoration for at least 10 years
- Conducting a long-term maintenance program
- Establishing funding sources for long-term maintenance and monitoring

Several important points are evident from this example, including not only creating restored wetlands at a 2:1 ratio to those lost, but also ensuring long-term success, monitoring, and maintenance. Establishing a means for funding of long-term monitoring and maintenance commitments is critical to any mitigation program.

Creation of new wetlands.   Off-site wetland creation normally involves converting an existing upland area to the type of wetland impacted or lost due to the proposed project or action. The design of a mitigation, or compensatory, wetland requires the coordinated efforts of hydrologists, geologists, engineers, and biologists (Fig. 20.6).

The first step is to locate a suitable site for the creation of replacement wetland. If a site is not available on or adjacent to the site of the proposed project, then off-site locations will need to be considered. Site selection should be based on characteristics that will enhance the probability of self-sustaining wetland functions; restore degraded wetlands if possible, as opposed to using upland areas; avoid complex hydraulic engineering features and/or questionable water sources that may create greater costs and a higher risk of failure in the replacement wetland.

The next step is to design the physical characteristics of the replacement wetland. After testing of soils and geology, the hydrology of the new wetland must receive careful consideration. Required procedures, such as possible use of clay liners to prevent immediate percolation of water into porous soils, must be identified in detail to ensure long-term maintenance of wetland hydrologic conditions. Detailed engineering construction plans and specifications are developed to indicate the required grading, excavation, soil replacement or liner, location of weirs or dams, etc.

The next step is an assessment of wetland vegetation species to be planted. This selection may be geared to match existing wetland vege-

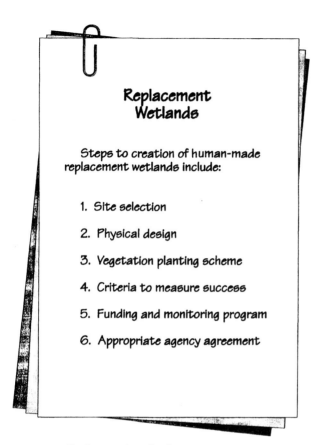

Replacement Wetlands

Steps to creation of human-made replacement wetlands include:

1. Site selection

2. Physical design

3. Vegetation planting scheme

4. Criteria to measure success

5. Funding and monitoring program

6. Appropriate agency agreement

Figure 20.6   Replacement wetlands.

tation in surrounding areas to ensure survival percentages. It may include special vegetative species that may enhance the ecosystem by adding more diversity to the vegetative community. The entire plan should reflect the type and chemical, physical, and biological functional values of the lost wetland. Therefore, vegetation selection may be based on wildlife food, cover, or nesting sites, or on ability to effectively trap sediment and improve water quality through removal of contaminants. A vegetation plan is then developed with the grading plan to indicate what species should be planted at what locations, by groupings. A materials list of required wetland vegetation is prepared.

The wetland replacement plan must include criteria by which to measure success and a funding and monitoring program.

Mitigation wetland banking.   Mitigation wetland banking is permitted under certain conditions (U.S. Department of Defense, Army Corps of

Engineers, EPA, NRCS, FWS, and NOAA 1995). *Mitigation banking* means the restoration, creation, enhancement, and in some cases preservation of wetlands expressly for the purpose of compensating future wetlands losses. A wetlands bank is created by restoring degraded wetlands or constructing new wetlands. A functional assessment methodology is used to quantify the amount of credits available for use from the bank. These credits are then subtracted from the bank to be used to offset wetland and aquatic resource impacts for particular projects or actions of the sponsoring agency.

The majority of mitigation banks established to date have been developed by state highway agencies or port authorities to compensate for future wetland impacts. Any private individual or organization, or local communities and counties, however, also may establish a mitigation bank.

20.8   Final Environmental Impact Statement

If a selected project or action alternative involves a wetland impact that cannot be avoided, the FEIS must include a Wetland Finding, or Only Practicable Alternative Finding, in compliance with Executive Order 11990 that includes

1. A reference to Executive Order 11990
2. An explanation of why there are no practicable alternatives to the proposed action
3. Documentation that the proposed action includes all practicable measures to minimize harm to wetlands.

Concurrence with this conclusion from all jurisdictional and interested resource agencies should be documented in the Final Environmental Impact Statement.

The FEIS also should note if a Section 404 permit, either individual or nationwide, will be required (refer to discussion in Chap. 18). Acres of wetlands to be lost should be specified, and firm commitments to specific mitigation should be made in the FEIS, including documentation of agreement by jurisdictional agencies on the developed mitigation plan. The mitigation plan should include identification of responsible agencies, document that funding is in place, and describe subsequent project monitoring programs.

Chapter

21

Vegetation
and Wildlife

The assessment of vegetation and wildlife impacts will likely be most
intense in rural areas and for proposed projects or actions covering
large geographic areas or setting future management policies. In
urban areas, however, small tracts of natural vegetation and habitat
may be extremely important if there is an absence of similar habitat
in the area or region. Wetlands are a significant habitat for numerous
species of plants and animals. See Chap. 20 for a discussion of wetland analysis.

Vegetation and wildlife studies often focus on threatened or endangered species. This focus is appropriate, particularly because such
species are protected by law. Studies must be conducted in a manner
to comply with the regulations, or project schedules can suffer. Game
species also will often receive more intense study because, among
other benefits, fishing and hunting can be a significant beneficial factor in state, regional, or local economies.

The environmental analyst should not, however, overlook the assessment of possible project or action effects on vegetative ecosystems
and wildlife species that neither are protected by law nor are game
species.

21.1 Describing Existing Resources

Vegetation and wildlife studies begin, as most other studies, with coordination with federal, state, and local agencies for information on
the presence of any special species or particularly valuable vegetation

types in the project area. Goals and objectives for the area should be reviewed, particularly if the project involves a large geographic area. Important contacts include the U.S. Fish and Wildlife Service, National Marine Fisheries Service, state natural resources or fish and wildlife agencies, and local planning agencies and organizations (such as fishing and boating groups).

For small, simple projects or actions, it will be sufficient to verbally describe the existing resources. As the level of expected impact increases, photographs and vegetation (habitat) mapping will most likely be required. Detailed studies are usually contained within a supporting technical report to the Draft Environmental Impact Statement or the Environmental Assessment. Habitat mapping can begin with a review of aerial photographs of the project area. Much preliminary work can be accomplished prior to doing any field surveys. Field surveys will then verify the habitat mapping and finalize the classification of vegetation communities.

Vegetative communities can be described generally or in terms of dominant species. Significant secondary species and understory species complete the description. Any special wildlife habitat features, such as feeding or nesting sites, water supplies, cover, or travel corridors, should be individually identified and emphasized. Unique or rare habitats or vegetative communities, relative to the presence of similar habitat types in the area or region, should be noted.

These are examples of the types of general vegetative community or wildlife habitat descriptors that may be used:

- *Hardwood forest*—areas where greater than 50 percent of the area is dominated by trees
- *Abandoned field scrub*—areas not subject to mowing for at least the current growing season and subject to invasion of woody plants
- *Agricultural*—areas maintained for annual crop production or pasturing; includes hedgerows and drainage ways
- *Human-dominated*—mowed aprons, lawns, and residential landscaping and gardens

For each of these general descriptors, supporting text would further describe the resources, including representative plant species. For example, the description of hardwood forest should include dominant species, understory species, and a discussion of tree size and forest successional maturity. For all natural areas, the extent of evidence of disturbance or intervention by humans may be important to note.

Examples of more detailed habitat descriptors may include

- Rivers, streams, floodplains, and wetlands
- Open water
- Marine and coastal areas
- Aquatic bed
- Riparian (streamside)
- Wetlands, by classification
- Sage scrub
- Scrub and shrub
- Annual grassland
- Oak-hickory forest
- Maple and beech forest
- Southern sycamore woodland
- Willow forest
- Conifer forest
- Bottomland hardwoods
- Ruderal (disturbed by humans)
- Ornamental and agricultural
- Developed, or urban

The environmental analyst should become knowledgeable about the communities of fish and wildlife, including reptiles and amphibians, that would be expected to be present within particular vegetative ecosystems or habitat types in the project area and region. It is not necessary to try to list all possible faunal (animal) species within the technical report or draft environmental document. Most often, such an attempt will not be complete. Examples of common species should be given, however, and any special species of concern definitely should be emphasized. Threatened and endangered species and state-listed species are discussed later in this chapter.

If the environmental impact assessment is being conducted on a large management plan, such as for a national forest or a Bureau of Land Management resource area, significantly more details will be involved in the description of existing vegetation and wildlife resources, the assessment of impacts of various management practices, and the selection of indicator species. For example, vegetative diversity in the Shawnee National Forest (U.S. Department of Agriculture, Forest Service 1992a) is assessed in terms of the following natural community descriptions:

1. Barrens
    a. Loess (dry-mesic) barrens
        (1) Cretaceous hills type
        (2) Shawnee hills type
        (3) Ozark hills type
    b. Limestone barrens
        (1) Ozark hills type
        (2) Shawnee hills type
    c. Sandstone barrens
    d. Sand barrens
    e. Cretaceous hills gravel barrens
    f. Ozark hills gravel barrens
2. Cliffs
    a. Sandstone cliffs
    b. Sandstone overhang
    c. Limestone cliff
3. Forests
    a. Xeric* upland forest
    b. Dry upland forest
    c. Dry-mesic upland forest
    d. Mesic upland forest
    e. Mesic floodplain forest
    f. Wet-mesic floodplain forest
    g. Wet floodplain forest
4. Woodlands—dry woodland
5. Wetlands
    a. Acid gravel seep
    b. Springs
    c. Swamps
    d. Shrub swamp
    e. Pond

Characteristic plant species are listed for each of the 25 identified natural communities. Detailed vegetative diversity and wildlife analyses are conducted, using computer models for incorporation of suitable habitat and population indices information.

A major tool in national forest management is the management indicator species (MIS). These species are selected to estimate the effects of forest management activities on wildlife communities and on the forest ecosystem as a whole. Each selected species is representa-

---

*Xeric plants require only a small amount of moisture, mesic plants require a moderate amount of moisture, and hydric plants require an abundant amount of moisture.

tive of a group (guild) of many other species that have the same general habitat requirements. Effects of management activities on the indicator species are assumed to represent the effects on other species in the guild. A forest plan will include a list of indicator species and calculation of acres of suitable habitat for each species.

21.2 Impact Analysis

The level of detail required for impact analysis for vegetation and wildlife will depend on the specific characteristics of the proposed project alternatives and the expected degree of effects. Examples of the types of impacts that may be applicable are given in this section, including loss of unique vegetative communities, direct loss of wildlife habitat and species, deterioration of remaining habitat, barriers to wildlife travel corridors, and effects on recreational activities and land use.

21.2.1 Loss of valuable vegetative community types

The analysis of degree of impact of direct loss of vegetation will depend heavily on the value of the vegetative community to be destroyed. If the vegetation is common and unremarkable, the effects can be quantified by amount of each type of community to be destroyed for each proposed alternative.

The key to ensuring an efficient analysis is the identification and quantification of any special or unique natural communities to be destroyed. Special areas would meet criteria such as rare in the area (such as virgin or mature forests), unique, or of high-quality functional value, such as wildlife habitat, erosion control, recreational use, or visual quality.

21.2.2 Direct loss of wildlife and habitat

For projects or actions requiring land clearance and removal of natural vegetation, the most obvious impact on wildlife will be loss of habitat and individual animals. Species with small home ranges will be most affected. Larger species may emigrate to adjacent areas, but the wildlife biologist should be cautious in assuming that adjacent areas can support any individuals that may invade. Often, the community will already be at its carrying capacity for the particular species, that is, at its maximum ability to support a particular number of individuals without causing stress or imbalance to the species population as a whole.

The amount of habitat, by type, destroyed by each proposed alternative should be quantified. Any special functions provided by the habitat, such as food supply, water supply, and nesting or resting resources, should be identified. Represented species that would incur loss of individuals also should be identified. There may be more emphasis placed on game species as a result of indicated agency, organization, or public interest and concern during the scoping process.

As with wetlands, methodologies exist for the assessment of the functional value of wildlife habitat, based on the number of functions provided and the quality and rarity of similar areas in the region. The analysis of vegetation and wildlife impacts should identify the functional attributes to be lost.

Two methods in common use for impact assessment are the habitat evaluation procedures (HEP) and the instream flow incremental methodology (IFIM) developed by the U.S. Fish and Wildlife Service (1981). Most methodologies use selected species as indicators for assessing potential impacts. These "evaluation species" should be selected to include species that are economically and socially important with a high degree of public interest, as well as species to indicate ecological conditions, such as the guild species of the U.S. Forest Service methodology. The selection of evaluation species will depend on the magnitude and type of expected impact and the characteristics and use of the affected area, that is, primary value for public use versus primary value for ecosystem features.

The Fish and Wildlife service mitigation policy (U.S. Department of Interior, Fish and Wildlife Service 1981) recognizes that a change in wildlife productivity or ecosystem structure and function may *not* result in a biologically adverse impact. The determination as to whether a biological change constitutes an adverse impact depends on the predicted future biological conditions with and without the proposed project or action.

The policy (U.S. Department of Interior, FWS 1981) defines *loss* requiring mitigation as

> A change in fish and wildlife resources due to human activities that is considered adverse and:
>
> (1) Reduces the biological value of that habitat for evaluation species;
>
> (2) Reduces population numbers of evaluation species;
>
> (3) Increases population numbers of "nuisance" species;
>
> (4) Reduces the human use of those fish and wildlife resources; or
>
> (5) Disrupts ecosystem structure and function.

Loss of habitat or individuals of a species for management purposes is not included as losses in the policy.

21.2.3   Deterioration of remaining habitat

An overall degradation of the remaining habitat may occur in a variety of ways. If the habitat destroyed provides an essential functional portion of a species' overall habitat requirements, such as nesting or feeding, the remainder of the habitat area will be incomplete in meeting the life requirements of the species and an overall loss in population numbers and characteristics could be expected.

Loss of even a small area of habitat of nonexceptional value may cause much more significant effects on wildlife populations if the loss results in the remaining habitat being split into smaller noncontiguous parcels. Species requiring large integral home ranges will be most affected.

Similarly, direct loss of a substantial number of individuals of a particular species may not be considered significant. If that loss affects the overall ecological, or food web, balance of the remaining habitat, there may be an adverse impact on numerous wildlife species occupying the remaining habitat. Such an impact may be particularly critical if the balance of predator and prey populations is upset.

Another example of a secondary type of impact that may occur is changes in the light exposure or moisture content of soils, which would, in turn, cause changes in the composition and diversity of the vegetative community and the wildlife populations.

Management proposals can affect vegetation and wildlife communities through such activities as permitting grazing which could increase soil erosion and deteriorate watersheds or remove important forest undergrowth, upsetting the normal successional progression. Management techniques designed to maintain large populations of economically valued game species can result in severe degradation of vegetative diversity and quality of habitat for the other species in the area.

Direct loss-of-cover impacts are immediate and permanent. Habitat degradation impacts, direct and secondary, and resultant effects on wildlife may be short-term if fairly rapid recovery is expected. The analysis of potential impacts on vegetation and wildlife also should attempt to conclude whether a proposed project or action would produce any long-term serious changes in the overall species' population levels or characteristics.

21.2.4   Barriers to travel corridors

The analyst must consider whether the proposed project or action would cause removal of connecting travel corridors between areas of wildlife habitat. Such corridors may cover large areas or may be very small but very important. This type of impact can often occur in suburban or partially rural areas where farmland or naturally vegetated land is being converted to developed use in a piecemeal approach. Often large natural areas may remain, but travelways for wildlife among the remaining large areas may be limited to narrow strips of woodland, fence rows, or riparian areas along streams.

Linear projects, such as highways, railroads, power lines, pipelines, or artificial drainage channels, are particularly likely to produce barriers to wildlife travel. The effect on wildlife populations can be particularly adverse, for example, if feeding or watering areas are separated from nesting or resting areas. Species requiring large home ranges are most affected. The analysis must consider not only the proportion of habitat lost, which in some cases may be a small percentage, but also whether that portion would split and render useless the remaining habitat because travel is restricted. Some species are more sensitive than others. For example, even a small two-lane roadway may present a genuine barrier if the particular species will not cross the paved area or is particularly sensitive to any human disturbance whatsoever. Some species require very remote areas.

21.2.5 Recreational use and enjoyment

Direct or indirect impacts to vegetation can produce secondary effects on recreational resource values. Vegetation is a major amenity in both expansive natural settings and smaller urban parks and open areas. Very small parks and open areas in urban environments can support recreational activities such as birding, picnicking, walking, bicycling, and general high-quality visual resources. Larger natural areas support fishing, camping, hunting, hiking, and research studies. To the degree possible, the impact of the proposed project or action alternatives on both active and passive recreational activities and qualities should be comparatively assessed and quantified.

21.3 Mitigation

Mitigation for potential impacts on vegetation and wildlife may be very site-specific, such as replacing landscaping or creating open space and parks in more urban areas; or geographically expansive in scope, such as implementing particular management techniques in national forests. In some cases, rare plants or particular animals may actually be transplanted or trapped and moved to other locations out of harm's way.

Mitigation measures to be considered are basically the same as those discussed for wetlands, and they follow the mitigation guidance of the CEQ Regulations, to include (1) avoiding, (2) minimizing, (3) rectifying, (4) reducing, and (5) compensating.

If particularly sensitive or valuable natural areas would be destroyed, the first mitigation technique should be development and feasibility analysis of avoidance alternatives. If total avoidance is not possible, design refinements may reduce the quantity of exceptional natural area affected.

Techniques to improve the productivity and functional value of the remaining habitat can be used to offset adverse impacts. Such measures may include installation of nesting boxes or trees, creation of waterholes and open spaces, plantings of food supply vegetation, or increasing the overall vegetational diversity.

Habitat impacts also are mitigated through compensatory preservation or created, replacement habitats. Depending on the value of the lost habitat, the required replacement ratio may be as high as 5:1, and the replacement should be functionally in kind to that lost. As with wetland mitigation programs, the habitat replacement plan should include detailed plans for physical construction and for plantings of various plant species. Sometimes, trees and vegetation removed by the proposed project or action can be saved and used to replant the created habitat.

Barriers to wildlife travel corridors can sometimes be mitigated through provision of wildlife underpasses in highway or railroads fills or similar types of protected travel corridors for power lines or artificial drainage channels. Wildlife losses through road kills can be further minimized by installation of fencing to prevent wildlife from crossing the highway and to direct wildlife movement to the provided underpasses.

Proposed mitigation measures should be coordinated with appropriate federal, state, and local agencies. The mitigation plan should include documentation of designated funding, responsible parties, performance criteria, monitoring methods, and schedule.

21.4 Endangered and Threatened Species

This section discusses the required compliance procedures for consideration of federally listed or proposed threatened and endangered species (Fig. 21.1). Many states also protect wildlife species through specific laws and regulations, as discussed next.

21.4.1 Legislation

The U.S. Fish and Wildlife Service is authorized under the Endangered Species Act of 1972 to establish lists of endangered and threatened plants and animals and to identify critical habitats for listed species. The lists are contained in 50 CFR Parts 17.11 and 17.12 and are published periodically in the *Federal Register.* Critical habitat is designated under the act and consists of that part of the geographic

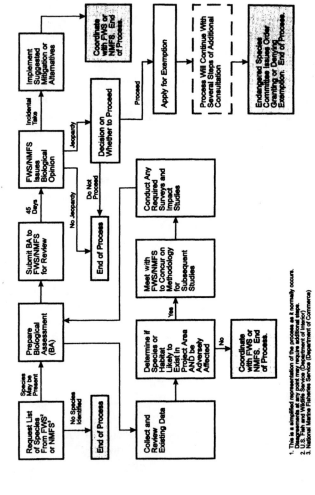

Figure 21.1 Endangered Species Act process.

1. This is a simplified representation of the process as it normally occurs. Disagreements at any point may require additional steps.
2. U.S. Fish and Wildlife Service (Department of Interior)
3. National Marine Fisheries Service (Department of Commerce)

area occupied by a listed species which contains those features essential to the conservation of the species. The National Marine Fisheries Service (NMFS) of the Department of Commerce also administers the Endangered Species Act and is responsible for protecting listed marine species. In this discussion, the two agencies together will be referred to as FWS/NMFS.

The act requires federal agencies to ensure that any action authorized, funded, or carried out by such an agency is not likely to jeopardize the continued existence of any endangered or threatened species or to result in the destruction or adverse modification of its designated critical habitat.

21.4.2   Identification of protected species

The initial step in compliance with the Endangered Species Act is a request to either the U.S. Fish and Wildlife Service or the National Marine Fisheries Service for a list of any listed or proposed species or designated habitat that may be present in the area of the proposed project or action. If a listed species or critical habitat may be present within the impact area of the proposed project or action, a biological assessment is required under Section 7 of the act.

21.4.3   Biological assessment

The biological assessment contents are at the discretion of the federal agency sponsoring the proposed project or action, but may include

- Results of on-site inspections or surveys
- Views of recognized experts
- Review of literature and other information
- Analysis of effects of the proposed project or action on the species and habitat
- Analysis of alternative actions considered

The biological assessment should be conducted in a level of detail suitable to the project or action characteristics and the biological requirements of the listed species. Keep in mind that the list received from FWS/NMFS will indicate species or habitat that *may* exist in the area. The list will usually encompass a very large geographic area, sometimes even all the species known, or thought, to occur in the entire state or county, even though the particular proposed project or action may affect only a very small area. Such a comprehensive approach is entirely appropriate. It remains the ultimate responsibility of the federal agency not to assist or sponsor any activity that may adversely affect an endangered species in compliance with the Endangered Species Act. The agency must therefore assume a proper share of accountability in identification of the presence of a listed species or critical habitat within the area of likely project effect.

In many circumstances, the biological assessment will be simple and obvious. For other proposed projects or actions, detailed surveys and analysis may be required. Such detailed studies, if required, can be extremely time-consuming and can have severe ramifications to the project or action implementation schedule.

First answer these three questions (Fig. 21.2):

1. Are there any historic sightings of the species within the project area? Review the literature and check with local universities and experts.

2. Does designated critical habitat exist in the project area? If not, does habitat required, or suitable for use by the species, for nest-

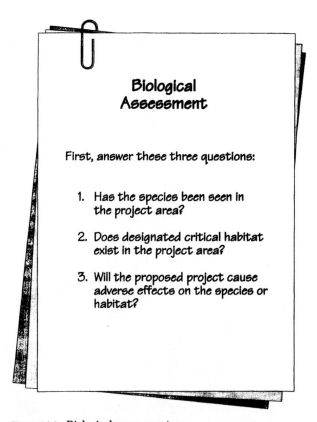

Figure 21.2 Biological assessment.

ing, feeding, or resting (animals) or survival (plants) exist in the project area? In some cases, the answer may be obvious. In other circumstances, a field view of the area of potential project effect may be required.

3. Will the characteristics of the project or action cause any disturbance or other adverse effects on such species or habitat known or thought to exist in the project area?

If the answers to the three questions above are all no, then a letter with documentation sent to the FWS/NMFS will normally result in a biological opinion that no adverse effects will occur, and the Endangered Species Act process is complete. The Draft Environmental Impact Statement or Environmental Assessment should include documentation of coordination with the FWS/NMFS and agreement on conclusions.

For projects requiring more in-depth analysis, it is extremely important to coordinate with the FWS/NMFS prior to conducting any detailed studies related to the biological assessment, to reach agreement and understanding on exactly what will be done. A detailed methodology, or protocol, should be developed in consultation with FWS/NMFS, especially for studies of species requiring field sampling and surveys. All parties should agree, in writing, to the approach, duration, and level of analysis detail of studies for each affected species before any work commences.

Previously prepared biological assessments for other projects in the area may be used if the information is verified to be current and still applicable. The biological assessment should be completed within 180 days after its initiation.

If the biological assessment concludes that there are no listed species or critical habitat present likely to be adversely affected by the action, and the FWS/NMFS concurs, the process is completed.

21.4.4 Formal consultation

If it is determined that the proposed project or action will affect a listed endangered or threatened species or critical habitat, formal consultation is initiated with the FWS/NMFS, with submission of the following information:

- Description of the proposed project or action
- Description of any listed species or critical habitat that may be affected
- Description of the effects on the species or habitat, including an analysis or any cumulative effects

- Relevant reports, including the Environmental Impact Statement or biological assessment prepared
- Any other relevant information

The formal consultation period can last up to 90 days. Within 45 days after completion of the formal consultation, the FWS/NMFS will issue a biological opinion as to whether the action, taken together with cumulative effects, is likely to jeopardize the continued existence of listed species or result in the destruction or adverse modification of critical habitat.

21.4.5 Biological opinion

A *no-jeopardy* biological opinion ends the process. A *jeopardy* biological opinion must include reasonable and prudent alternatives that the agency can take to avoid violation of the act. If the FWS/NMFS cannot develop such alternatives, the biological opinion must indicate that there are no reasonable and prudent alternatives.

If a project would affect a listed species or critical habitat, but not to the jeopardy extent protected by the Endangered Species Act, the impact is referred to as an *incidental take*. The biological opinion from the FWS/NMFS for an incidental take will specify

- The impact
- Reasonable and prudent mitigation measures
- Terms and conditions to implement the mitigation measures
- The procedures to be used to handle or dispose of any individuals of a species actually taken

An example of such mitigation is the relocation of individual plants or animals to other areas.

If the sponsoring agency decides to proceed with the proposed project or action after a jeopardy biological opinion is issued, the agency must inform the FWS/NMFS and apply for an exception to the Endangered Species Act.

21.4.6 Incidental take permit for nonfederal projects

The 1982 amendments to the Endangered Species Act included a provision to allow incidental takes of endangered and threatened species of wildlife by nonfederal entities. Federal projects and nonfederal projects with federal funding or permitting obtain incidental take authority through the consultation process under Section 7 of the act, as discussed above. The incidental take permit is issued by the U.S. Fish and Wildlife Service and is also referred to as a Section 10 permit or, more specifically, as a Section 10(a)(1)(B) permit.

The permit process (U.S. Department of Interior, FWS 1995) requires preparation of a habitat conservation plan (HCP). The HCP must include the following elements:

- Impacts, including anticipated take levels in terms of number of individual plants or animals or in terms of habitat acres
- Mitigation programs and standards, including any compensation in the form of habitat banking or replacement
- Monitoring measures
- Unforeseen circumstances and plan amendments
- Funding
- Alternatives analyzed
- Implementing agreements

The project sponsor must apply for a Section 10 permit through submission of a complete package to FWS, including the HCP. The process and review time varies depending on the category of conservation planning efforts and permit applications as high-effect, medium-effect, or low-effect. The FWS has specific permit issuance criteria that are assessed in determining whether to issue the permit.

21.5 State-Protected Species

Many states have legislation to protect important or special plant and animal species. Early coordination with appropriate state and regional agencies should include a request for information on any species of special concern that may exist within the area of the proposed project or action. The study team should become thoroughly familiar with the accepted procedures and processing required for compliance with state regulations and guidelines. In some cases, the amount of study required for state-listed species may be more detailed than that required for compliance with the federal Endangered Species Act. State regulations also would apply to projects with no involvement of federal assistance or funding.

For example, the state of California has an Environmental Quality Act, an Endangered Species Act, a Native Plant Protection Act, a Wild and Scenic Rivers Act, and codes requiring "streambed alteration agreements" for modifications that may affect fish and wildlife resources. Environmental impact assessments in California must consider species listed on the federal FWS/NMFS list; species listed as candidate for federal listing; species listed as endangered, threatened, or rare by the state; California Department of Fish and Game species of special concern; federal sensitive species listed by the Bureau of Land Management and U.S. Forest Service; and species listed by the California Native Plant Society. Databases available include the California Natural Diversity Data Base and the Wildlife Habitat Relationships Data Base. Specific forms have been developed for use in the field for wetland surveys, native-species field surveys, and natural community field surveys.

21.6 Contents of Environmental Document

If the required wildlife and vegetation studies are detailed and extensive, the description of study methodologies, sampling or surveys, analysis, and results should be contained within a technical report supporting the Environmental Assessment or Draft Environmental Impact Statement. The environmental document should summarize the information from the technical report in a comparative manner for the proposed alternative, including the no-build alternative. Impacts should be quantified to the degree possible; mitigation opportunity and expected success should be comparatively stated for each alternative.

Documentation with appropriate federal, state, and local agencies should be provided, including an opinion on the identification and value of existing resources, conclusions of the impact analysis, and proposed mitigation measures.

All specific legislative requirements for public and agency review should be met within the environmental document, if possible.

22

Comparative Evaluation

Following completion of all appropriate environmental impact assessment studies, the major task is to make the completed analyses productive to the decision-making process. Although this statement sounds quite obvious, many environmental impact assessments flounder at this final stage of actually using results productively. The evaluation of alternatives must result in a clear, concise comparison that easily illustrates the tradeoffs involved between the build and no-build alternatives and the distinguishing degree of impact among the various build, or action, alternatives. A recommended process for evaluation of alternatives is illustrated in Fig. 22.1.

22.1 Evaluation Criteria

The development of evaluation criteria should actually take place prior to conduct of the environmental studies (refer to Chap. 4, Alternatives). The establishment of what will be used to judge the considered alternatives early in the study phase adds credibility to the decision-making process by eliminating the possibility that criteria will be developed to specifically favor or benefit a particular alternative.

Evaluation criteria should be based on the identified purpose and need for the proposed project or action and the goals and objectives (refer to Chap. 4). The evaluation criteria advance the goals and objectives process one step by establishing the parameters by which each impact will be assessed. The parameters also measure the ability of each alternative to meet identified goals and objectives. For example, the measurement parameters for comparison of alternatives for a typical highway project may include these:

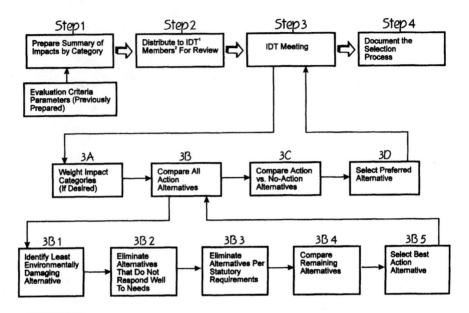

Figure 22.1 Evaluation process.

Goal: Reduce traffic congestion

Measurement parameters
- Number of highway segments operating at level of service D or worse
- Calculated duration of delay at signals

Goal: Ability to be implemented

Measurement parameters
- Costs (compared with available funding)
- Degree of controversy
- Length of construction period

Goal: Minimize noise impact

Measurement parameter
- Length of required noise wall

Goal: Minimize natural resource impacts

Measurement parameters
- Acres of wetland affected

- Floodplain encroachment (yes or no)
- Endangered species involvement (yes or no)
- Violation of water quality standards (yes or no)
- Acres of wildlife habitat destroyed

Goal: Minimize socioeconomic effects

Measurement parameters
- Number of jobs created
- Number of required business and residential relocations
- Loss of tax base (in dollars)
- Number of community facilities adversely affected
- Creation of a barrier to neighborhood cohesion (yes or no)

Goal: Avoid cultural effects

Measurement parameters
- Adverse effect on National Register listed historic site (yes or no)

The above examples are certainly not all-inclusive, but they give an idea of the type of parameters to be developed for each area of potential impact. Measurement parameters should be quantitative if possible. In some areas of impact, however, such as visual effects, this may not be easy. Visual effects often are very subjective, but the comparison should be as objective as possible. Although there are methodologies to produce numerical results from a visual impact analysis (refer to Chap. 13), these numbers are usually generated through subjective assessment and thus have limited value. Often with impact areas such as visual resources, a short description of the amount of physical change will be required for each considered alternative. In this and some other areas of impact, such as social or community effects, it may be necessary just to use *minimal, moderate,* or *severe* to describe the comparative effects of the alternatives.

The goal is to present the results of environmental analyses as objectively as possible and to let the reader consider and comparatively weigh the benefits versus disadvantages of each proposed alternative.

22.2 Summary of Analysis Results

The next step in evaluation is to compile summaries of the results of all conducted environmental (and engineering, if applicable) studies. If a technical report was prepared for a particular area of impact, it should always contain a summary. That summary can now be used for evaluation.

To be effective, the summary should be as concise as possible. It is extremely important that impact summaries be written in a manner understood by laypersons outside the particular discipline and by the general public.

The summary should be reviewed and key elements converted to a form compatible with the established evaluation criteria measurement parameters. Although the evaluation will be presented, usually in tabular form, in the summary section of the Environmental Assessment or Draft Environmental Impact Statement, it will be extremely concise in that format. It is often a good idea to prepare an additional, somewhat longer summary of environmental impacts for use as a handout at the public hearing or meeting. Tables should be used as much as possible in the handout, but impact areas requiring explanatory narrative also can be presented in a few brief paragraphs.

The Environmental Assessment or Draft Environmental Impact Statement normally should not identify a recommended, or preferred, alternative. If, in fact, a particular alternative is indeed preferred at this stage of studies, however, it should be identified as such in the draft environmental document.

22.3   Weighting Impact Categories and Interpreting Significance

It is generally accepted that not all possible impact categories* have equal importance when one is considering an evaluation of tradeoffs among alternatives. Some areas of potential effect receive great importance because of protection by legislation and regulations, such as wetlands, historic sites, floodplains, and endangered species. Granting these areas of potential effect greater importance than other categories is appropriate because the very existence of protective regulations indicates a national level of concern and degree of importance of such resources.

The relative importance of other impact categories may be very individual to the particular proposed project or action characteristics or to the existing features of the geographic region or site. The public and agency scoping process is designed to identify those resources and potential impacts considered most important based on the possible effects of the particular proposed project or action and the characteristics of the region or local setting. The results of scoping can now be used to include community and agency opinion in the process of comparative evaluation of alternatives.

---

*Impact categories refer to the various disciplinary areas of potential impacts investigated, such as air quality, noise, land use, wetlands, economic effects, and historic sites.

Several agencies and states have developed methodologies for establishing the relative importance of impact categories and interpreting the significance of impacts. Chapter 1 of this text discussed the criteria of significance contained within the Council on Environmental Quality regulations, referred to as criteria to measure severity, or intensity, of project or action effect on the human environment. These criteria include impact category elements, such as historic sites, endangered species, and public health, and project-related criteria, such as degree of controversy, setting an irreversible precedent for future actions (limiting the options available to future generations), and the degree of uncertainty in predicting effects.

One of the earliest developed methodologies for evaluating environmental impacts was a detailed matrix system developed by the USGS (U.S. Department of Interior, 1971). The matrix includes a list of 100 actions which cause environmental impact on one axis and a list of 88 existing environmental conditions that may be affected on the other axis, which gives a total of 8800 possible interactions. Numerical importance is assigned to each possible relevant interaction for a particular proposed project based on magnitude and importance.

Although titled an "evaluation" procedure, the purpose stated in the USGS publication is actually most related to "reminding investigators of the variety of interactions that might be involved" and identifying alternatives which might lessen impact. Matrix rows or columns with numerous interactions should be covered in detail in the environmental analysis and document. Study team members today will most likely find the assignment of numerical values tedious and of minimal use. Considering the time of this publication, however, when those charged with assessing environmental impacts in compliance with the "new" National Environmental Policy Act had no prior Environmental Impact Statements to review and minimal agency, regulatory, or legislative guidance, this publication played an important role in exhibiting a full range of possible impacts that may apply to any particular proposed project or action. The matrix today is still valuable for use in scoping and mitigation analysis.

Another relatively early publication, the 1977 Soil Conservation Service's *Guide for Environmental Assessment,* focuses more on the environmental impact analysis process, preliminary matrix overviews to "scope" the study, and an example of a detailed checklist evaluation system to compare alternatives after studies are completed. This publication remains extremely useful today. The assessment study process is sound, and it includes the following steps:

1. Initial phase of assessment
   a. Counsel sponsors
   b. Identify study needs

   c. Establish interdisciplinary team
   d. Obtain resource data
   e. Make field examination
   f. Summarize initial assessment
2. Detailed phase of assessment
   a. Develop strategy
   b. Describe present conditions
   c. Predict future conditions
   d. Calculate impacts
   e. Summarize detailed assessment

The example evaluation checklist begins with assessment of land, water, and air resources and then presents seventeen individual resource use checklists. The methodology includes a good example of evaluation parameters and appropriate measurement units for comparison of future resource use conditions with and without the proposed project or action. In several impact categories, value rating systems using relative weightings are suggested to assist in quantifying impacts.

The U.S. Water Resources Council (1983) has published economic and environmental principles and guidelines to assist water resource development agencies in formulation and evaluation studies. The basic steps of the planning process are as follows:

1. Identification of problems and opportunities

2. Inventory and forecast of resource conditions

3. Formulation of alternative plans

4. Evaluation of effects

5. Comparison of alternative plans

6. Plan selection

The publication presents economic development evaluation procedures for benefits and costs. It also demonstrates an evaluation process for environmental quality.

One of the most detailed evaluation methodologies is that used by the Federal Transit Administration (FTA) to rank proposed projects for federal assistance (U.S. Department of Transportation, FTA 1993). Although included in the evaluation system, environmental factors are secondary to the primary emphasis on transportation service. A detailed financial and cost/benefit analysis is required, and indices of project merit are calculated. Proposed projects must then meet threshold or screening tests at three stages of the project development process. The major focus of the threshold tests is to establish minimum levels of cost effectiveness for federal funding assistance.

Review of the above examples will assist the environmental analyst in developing project-specific evaluation methodologies for the particular projects or actions being considered. The actual method, however, should include local factors and opinions of local and state agencies.

22.4   Selecting a Preferred Alternative

After the Environmental Assessment or Draft Environmental Impact Statement has been made available to the public and has been circulated to interested agencies, all comments received should be summarized. Any additional environmental analysis will be conducted, and considered alternatives will be reevaluated for possible changes to further minimize impacts or respond to comments received.

A revised summary of impacts of each alternative should be compared, using the evaluation criteria and measurement parameters. The next task is the selection of the preferred alternative. In some cases, the preferred alternative may be obvious, and the selection process will be brief. In other proposed projects or actions, a more thorough analysis and process will be required.

Regardless of the process used, it is important that it be systematic and logical. Documentation of decisions made and the reasons behind each decision should be prepared. The following is a good system to use for fairly involved projects or actions.

Each team member should prepare a brief summary of the impacts and comments received within his or her discipline (air quality, noise, social effects, wildlife, etc.). These summaries should be distributed to all team members for review. A meeting of all team members can then be held to discuss the pros and cons of each alternative in each area of potential impact.

A good approach is to compare the build, or action, alternatives first. The least environmentally damaging alternative, with mitigation in place, should be identified. If any considered build alternatives are clearly less responsive to the identified project purpose and need, they should be eliminated first. There is little sense in proceeding with a proposed project or action if it cannot accomplish the basic goals and objectives to meet established needs.

The next step is to compare the remaining build or action alternatives for legislative or regulatory restrictions. As discussed in Chaps. 5 through 21, numerous types of potential impacts are regulated by specific guidelines that would prohibit selection of a particular alternative under certain conditions, such as the existence of a feasible and prudent alternative, or a less-environmental-impact alternative, in the remaining set of alternatives. There also may be circumstances where a jurisdictional agency has indicated a future denial of a necessary permit for a particular alternative. Any alternative not meeting legislative or regulatory requirements must be eliminated from further consideration.

The remaining build or action alternatives are then compared in detail, including such criteria as opportunity for mitigation of adverse effects, project costs, severity of impact in any particular area, public and political opinions, and other established evaluation standards. Through interaction of the interdisciplinary team, an alternative is selected as the preferred action alternative.

The next step after the preferred build, or action, alternative is selected is to directly compare it with the no-build alternative. The team is now at the final stage of build versus no build. This is the

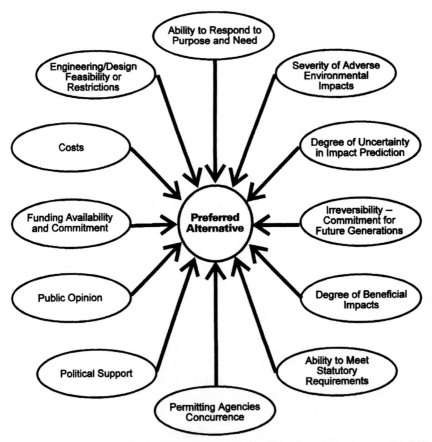

Figure 22.2 Factors affecting selection of a preferred alternative and ultimate project or action implementation. (Some factors often will be more important than others in the decision-making process.)

phase where tradeoffs should be clearly presented and evaluated. The analysis of benefits versus costs, with incorporation of any agency-specific feasibility criteria, will lead to the final decision on whether the identified preferred alternative is the selected build alternative or the no-action alternative.

With the selection of a preferred alternative and completion of the Final Environmental Impact Statement and Record of Decision, the environmental impact study and NEPA process is completed. Committed mitigation monitoring programs will continue with the project or action through construction. Other considerations may, however, still prevent the proposed project or action from proceeding to construction or implementation. A summary of the major factors which enter into the decision-making process for selection of a preferred alternative and for ultimate project completion is illustrated in Fig. 22.2.

Bibliography

Advisory Council on Historic Preservation. 1986a. *Section 106, Step-by-Step.*
——. 1986b. *Protection of Historic Properties.* 36 CFR Part 800.
——. 1993. *National Historic Preservation Act of 1966, as Amended,* 3d ed.
——. 1994. *A Five-Minute Look at Section 106 Review.*
Advisory Council on Historic Preservation, National Park Service, and U.S. Department of Interior. 1989. *The Section 110 Guidelines.*
California Department of Transportation. 1982. *Estimating Highway Runoff Quality.* Office of Transportation Research. FHWA/CA/TL-82/11. Sacramento.
——. 1988. *Conducting Socioeconomic Analysis.* Office of Environmental Analysis. Sacramento.
——. 1990a. *Procedures for Completing the Natural Resources Environmental Study and Related Biological Reports.* Office of Environmental Analysis. Sacramento.
——. 1990b. *Water Quality Technical Analysis Notes.* Division of New Technology, Materials, and Research. Prepared by A. E. Bates. Sacramento.
——. 1992. *Relocation Impact Documents.* Chapter 602-1. Sacramento.
——. 1995. *Local Program Manual, Environmental Procedures.* Sacramento.
——. 1995. *Environmental Handbook, Manual of Instructions.* Sacramento.
California Office of Planning and Research. 1973. *State CEQA Guidelines.* Title 14, Chapter 3, as amended in 1983.
California Public Resources Code. *California Environmental Quality Act.* Sections 21000–21177, as amended.
Council on Environmental Quality. 1978. *Regulations for Implementing the Procedural Provisions of the National Environmental Policy Act.* 40 CFR Parts 1500–1508.
——. 1980. *Procedures for Interagency Consultation to Avoid or Mitigate Adverse Effects on Rivers in the Nationwide Inventory.* Memorandum for heads of agencies. August 10.
——. 1981a. *Forty Most Asked Questions Concerning CEQ's National Environmental Policy Act Regulations.* 40 FR 18026.
——. 1981b. *Scoping Guidance.*
——. 1983. *Guidance regarding NEPA Regulations.* 48 FR 34263.
——. 1984. *Appendices to the CEQ Regulations.* 49 FR 49750.
——. 1989. *Inventory of Federal Agency Activities on Cumulative Impact Assessment and Summary of November 30, 1988 Interagency Meeting on Cumulative Impact Assessment.* January.
County Sanitation Districts of Los Angeles County. 1994. *Draft Program Environmental Impact Report for the Joint Outfall System 2010 Master Facilities Plan.* Whittier, Calif.
Federal Emergency Management Agency. 1986. *A Unified National Program for Floodplain Management.* March.
Interagency Working Group on Federal Wetlands Policy. 1993. *Fact Sheets.* August.
Lord, Byron. 1987. *Non-Point Source Pollution from Highway Stormwater Runoff.* Presented at Transportation Research Board conference, Washington, D.C. January.
Los Angeles County Metropolitan Transportation Authority. 1993. *Draft Supplemental Environmental Impact Statement, Metro Green Line Northern Extension.* Los Angeles.

Los Angeles County Transportation Commission. 1984. *Draft Environmental Impact Report, The Long Beach–Los Angeles Rail Transit Project*. Los Angeles.

Pennsylvania Department of Transportation. 1993. *The Transportation Project Development Process, Environmental Impact Statement Handbook*. Publication 278. Harrisburg.

——. 1994a. *Categorical Exclusion Evaluation for S.R. 1033, Hollow Road over French Creek, Chester County, Pa.* District 6-0, Philadelphia.

——. 1994b. *The Transportation Project Development Process, Categorical Exclusion Evaluation Handbook*. Publication 294. Harrisburg.

U.S. Department of Agriculture, Forest Service. 1992a. *Final Supplemental Environmental Impact Statement, Amended Land and Resource Management Plan, Shawnee National Forest*. Eastern region. Milwaukee.

——. 1992b. *Record of Decision, Final Supplemental Environmental Impact Statement, Amended Land and Resource Management Plan, Shawnee National Forest*. Eastern region. Milwaukee.

——. 1992c. *Record of Decision for Oil and Gas Leasing, Final Supplemental Environmental Impact Statement, Amended Land and Resource Management Plan, Shawnee National Forest*. Eastern region. Milwaukee.

——. 1992d. *Shawnee National Forest Amended Land and Resource Management Plan*. Eastern region. Milwaukee.

U.S. Department of Agriculture, Natural Resources Conservation Service. 1995. *Environmental Effects for Resource Management Plans*.

U.S. Department of Agriculture, Soil Conservation Service. 1977. *Guide for Environmental Assessment*.

——. 1983. *National Agricultural Land Evaluation and Site Assessment (LESA) Handbook*. 310-VI-NLESAH.

——. 1988. *National Food Security Act Manual*, 2d ed. Title 180. August.

——. 1990. *Materials in Support of Environmental Evaluation Activities*.

U.S. Department of Commerce, National Oceanic and Atmospheric Administration. 1985. *Federal Coastal Programs Review*. Office of Ocean and Coastal Resource Management.

U.S. Department of Defense, Army Corps of Engineers. 1979. *Wetland Values: Concepts and Methods for Wetlands Evaluation*. Institute for Water Resources. Research Report 79-RI.

——. 1987. *Wetland Delineation Manual*. Technical Report Y-87-1.

——. 1991. *Regulatory Programs of the Corps of Engineers; Final Rule*. 33 CFR Parts 320–330. Originally 51 FR 41206, November 13, 1986, as amended FR February 3, 1988, December 8, 1989, June 29, 1990, November 22, 1991, August 25, 1993, and March 14, 1995.

——. 1995. *Nationwide Permit for Single-Family Housing: Notice*. 60 FR 38650.

U.S. Department of Defense, Army Corps of Engineers, Los Angeles District. 1993. *Habitat Mitigation and Monitoring Proposal Guidelines*. Los Angeles.

U.S. Department of Defense, Army Corps of Engineers, and U.S. Department of Transportation Federal Highway Administration. 1987. *Wetland Evaluation Technique (WET)*, vol. 11. Methodology. FHWA-IP-88-029.

U.S. Department of Defense, Army Corps of Engineers, and U.S. Environmental Protection Agency. 1989a. *Memorandum of Agreement...concerning the Determination of the Geographical Jurisdiction of the Section 404 Program and the Application of the Exemptions under Section 404(f) of the Clean Water Act*.

——. 1989b. *Memorandum of Agreement...concerning Federal Enforcement for the Section 404 Program of the Clean Water Act*.

——. 1990. *Memorandum of Agreement...concerning the Determination of Mitigation under the Clean Water Act Section 404(b)(1) Guidelines*.

——. 1993. *Clean Water Act Regulatory Programs; Final Rule*. 33 CFR Parts 323 and 328, 40 CFR Part 110, etc.

——. 1995. *Individual Permit Flexibility for Small Landowners*.

U.S. Department of Defense, Army Corps of Engineers; U.S. Environmental Protection Agency; U.S. Department of Agriculture, Natural Resources Conservation Service; U.S. Department of Interior, Fish and Wildlife Service; and U.S. Department of Commerce,

National Oceanic and Atmospheric Administration. 1995. *Federal Guidance for the Establishment, Use and Operation of Mitigation Banks.* 60 FR 12286.

U.S. Department of Interior. 1993. *Departmental Manual, Part 516, National Environmental Policy Act of 1969, and Appendices.*

U.S. Department of Interior, Bureau of Land Management. 1984. *Visual Resource Management.* BLM Manual Section 8400.

———. 1986a. *Visual Resources Inventory.* H-8410-1.

———. 1986b. *Visual Resource Contrast Rating.* H-8431-1.

———. 1988. *National Environmental Policy Act Handbook.* H-1790-1.

U.S. Department of Interior, Bureau of Land Management, California Desert District, and County of Imperial, Calif. 1995. *Final Environmental Impact Statement and Environmental Impact Report, Mesquite Regional Landfill, Imperial County, CA.*

U.S. Department of Interior, Bureau of Land Management, California State Office. 1993. *Environmental Analysis Handbook.*

U.S. Department of Interior, Bureau of Land Management, Miles City District Office. 1995. *Final Big Dry Resource Management Plan/Environmental Impact Statement.*

U.S. Department of Interior, Fish and Wildlife Service. 1979. *Classification of Wetlands and Deepwater Habitats of the United States.* FWS/OBS-79/31. Office of Biological Services.

———. 1981. *U.S. Fish and Wildlife Service Mitigation Policy.* 46 FR 7643.

———. 1984. *NEPA Handbook.*

———. 1988. *National List of Plant Species that Occur in Wetlands: 1988 California (Region O).* Biological Report 88(26.10). Prepared by Porter B. Reed, Jr.

———. 1994. *Interagency Activities Environmental Review.* Part 505 FW 1–5.

———. 1995. *Preliminary Draft Handbook for Habitat Conservation Planning and Incidental Take Permit Processing.* September.

U.S. Department of Interior, Fish and Wildlife Service, and U.S. Department of Commerce, National Oceanic and Atmospheric Administration. 1986. *Interagency Cooperation—Endangered Species Act of 1973, as Amended; Final Rule.* 50 CFR Part 402.

U.S. Department of Interior, National Park Service. 1982. *How to Apply the National Register Criteria for Evaluation.*

———. 1983. *Archaeology and Historic Preservation; Secretary of Interior's Standard and Guidelines.* 48 FR 44716.

U.S. Department of Interior, National Park Service, and U.S. Department of Agriculture, Forest Service. 1982. *National Wild and Scenic Rivers System, Final Revised Guidelines for Eligibility, Classification and Management of River Areas.* 47 FR 39454.

U.S. Department of Interior, U.S. Geological Survey. 1971. *A Procedure for Evaluating Environmental Impact.* Prepared by L. B. Leopold, F. E. Clarke, B. B. Hanshaw, and J. R. Balsley. Geological Survey Circular 645. Thirteenth printing 1991.

———. 1979. *A Methodology for Post-EIS (Environmental Impact Statement) Monitoring.* Prepared by L. G. Marcus. Geological Survey Circular 782.

———. 1993. *Managing and Implementing the NEPA Process.*

———. 1994. *USGS Special Expertise on Environmental Quality Issues.*

———. 1995a. *Water—Managing a National Resource.* Fact Sheet FS-059-95.

———. 1995b. *Bioremediation: Nature's Way to a Cleaner Environment.* Fact Sheet FS-054-95.

———. 1995c. *Programs in the Great Lakes.* Fact Sheet FS-056-95.

———. 1995d. *Water Data Program.* Fact Sheet FS-065-95.

———. 1995e. *Ground-Water Studies.* Fact Sheet FS-058-95.

———. 1995f. *An Overview of the Stream-Gauging Program.* Fact Sheet FS-066-95.

———. 1995g. *National Water-Use Information Program.* Fact Sheet FS-057-95.

———. 1995h. *Northern California Storms and Floods of January, 1995.* Fact Sheet FS-062-95.

———. 1995i. *Programs in California.* Fact Sheet FS-005-95.

———. 1995j. *Chesapeake Bay: Measuring Pollution Reduction.* Fact Sheet FS-055-95.

U.S. Department of Justice, Bureau of Prisons. 1994. *Facilities Development Manual.*

———. 1995. *Final Environmental Impact Statement/Environmental Impact Report, Fort Devens, Massachusetts, Federal Medical Center Complex.* EOEA 9885/9116.

U.S. Department of Transportation, Federal Highway Administration. 1979. *Location and Hydraulic Design of Encroachment on Floodplains*. FHPM 6-7-3-2. 23 CFR Part 650.

——. 1982. *A Method for Wetland Functional Value Assessment, Volumes I and II*. FHWA-IP-82-23 and FHWA-IP-82-24.

——. 1985. *Effects of Highway Runoff on Receiving Waters*. Volume 4, Procedural Guidelines for Environmental Assessments.

——. 1986. *Visual Impact Assessment for Highway Projects*. Office of Environmental Policy.

——. 1987a. *Guidance for Preparing and Processing Environmental and Section 4(f) Documents*. FHWA Technical Advisory T 6640.8A.

——. 1987b. *Guidance on Cooperating Agencies*. Office of Environmental Policy.

——. 1987 and 1989. *Section 4(f) Policy Paper and Additions*.

——. 1987c. *Preparation of Environmental Documents*. NHI Course No. 14205.

——. 1988. *Environmental Impact and Related Procedures*. 52 FR 32646.

U.S. Department of Transportation, Federal Highway Administration, California Department of Transportation, Riverside County Transportation Commission, and San Bernardino Associated Governments. 1993. *Draft Technical Reports for I-215 Improvement Project: Land Use Impact Analysis, Relocation Impact Study, Socioeconomic Impact Analysis, Visual Impact Analysis, Energy Report, Air Quality Analysis, Floodplain Evaluation Report, Water Quality Report, Natural Environmental Study and Biological Assessment*. San Bernardino, Calif.

U.S. Department of Transportation, Federal Highway Administration, California Department of Transportation, and San Bernardino Associated Governments. 1993. *Draft Environmental Impact Statement, Widening of I-215 between I-10 and SR 30 in San Bernardino, CA., and Supporting Technical Reports: Noise, Air Quality, Traffic Analysis, Intersection Analysis Report, Water Quality Study, Natural Environment, Background Socioeconomic and Land Use Study, Initial Site Assessment*. San Bernardino, Calif.

——. 1994. *Final Environmental Impact Statement, Widening of I-215 between I-10 and SR 30 in San Bernardino, CA*. San Bernardino, Calif.

U.S. Department of Transportation, Federal Highway Administration, California Department of Transportation, and Transportation Corridor Agencies of Orange County. 1989. *Screencheck Environmental Impact Report/Environmental Impact Statement, The San Joaquin Transportation Corridor*. Orange, Calif.

U.S. Department of Transportation, Federal Highway Administration, and Maryland State Highway Administration. 1988. *Draft Environmental Impact Statement, U.S. Route 29, Montgomery County*. Baltimore, Md.

U.S. Department of Transportation, Federal Highway Administration, and Pennsylvania Department of Transportation. 1993. *Draft Environmental Impact Statement and Section 4(f) Evaluation, U.S. 222 Corridor Design Location Study, Lehigh County*.

——. 1995. *Environmental Assessment for the Hazleton Beltway Project*.

U.S. Department of Transportation, Federal Transit Administration. 1993. *Procedures and Technical Methods for Transit Project Planning*.

U.S. Department of Transportation, Federal Transit Administration, and Los Angeles County Transportation Commission. 1992. *Alternatives Analysis/Draft Environmental Impact Statement/Draft Environmental Impact Report, Eastside Corridor, Conceptual and Detailed Definition of Alternatives*.

U.S. Department of Transportation, Urban Mass Transportation Administration, and City and County of Honolulu, Department of Transportation Services. 1989. *Alternatives Analysis and Draft Environmental Impact Statement, Honolulu Rapid Transit Development Project, Technical Report on Noise and Vibration*.

——. 1990. *Alternatives Analysis and Draft Environmental Impact Statement, Honolulu Rapid Transit Development Project*.

U.S. Department of Transportation, Urban Mass Transportation Administration, and Metro-Dade County Transit Agency. 1987. *Draft Environmental Impact Statement, Miami Metromover*.

U.S. Environmental Protection Agency. 1980. *Guidelines for Specification of Disposal Sites for Dredged or Fill Material*. 40 CFR Part 230. 45 FR 85336.

———. 1984. *Policy and Procedures for the Review of Federal Actions Impacting the Environment.* Office of Federal Activities.

———. 1989. *Tools for Local Governments.* Office of Water. WH-550G. EPA 440/6-89-002.

———. 1991. *Cross-Cutting Environmental Laws, A Guide for Federal/State Project Officers.* Office of Federal Activities.

——— 1993. *The National Environmental Policy Act and Environmental Protection Agency Programs.* EPA Workgroup on NEPA.

———. 1994. *Draft Environmental Impact Statement, Eagle Pass Mine, Maverick County, Texas.*

———. 1995a. *Final Environmental Impact Statement, Eagle Pass Mine, Maverick County, Texas.*

———. 1995b. *Record of Decision, Eagle Pass Mine, Maverick County, Texas.*

———. 1995c. *EPA's Section 309 Review: The Clean Air Act and NEPA.* Office of Federal Activities.

———. 1995d. *Wetlands Fact Sheets.* EPA843-F-95-001.

———. 1995e. *Draft and Final Environmental Impact Statements for the Full Secondary Treatment Upgrade Project at the Joint Water Pollution Control Plant in Carson, CA.* Region IX, San Francisco.

U.S. Water Resources Council. 1978. *Guidelines for Implementing Executive Order 11988.*

———. 1983. *Economic and Environmental Principles and Guidelines for Water and Related Land Resources Implementation Studies.*

White House Office on Environmental Policy. 1993. *Protecting America's Wetlands: A Fair, Flexible, and Effective Approach.* August.

Index

ABOUT THE AUTHOR

Betty Bowers Marriott is an environmental consultant with 20 years of experience conducting and managing environmental impact assessments and preparing environmental documents. Specializing in transportation projects, she has worked throughout the United States on highway and transit projects involving the application of FHWA, FTA, and a multitude of state-specific environmental regulations and guidelines. Her experience with federal regulations includes Section 4(f), Section 106, Executive Orders protecting floodplains and wetlands, the Farmland Protection Policy Act, and accepted procedures and content for air quality and noise analyses. She is a member of the California Association of Environmental Professionals, and was previously a founding member of the Pennsylvania Association of Environmental Professionals.